Environmental Management Systems and Cleaner Production

Environmental Management Systems and Cleaner Production

Edited by
Ruth Hillary
*Centre for Environmental Technology,
Imperial College of Science, Technology and Medicine, UK*

JOHN WILEY & SONS
Chichester · New York · Weinheim · Brisbane · Singapore · Toronto

Copyright © 1997 by John Wiley & Sons Ltd,
Baffins Lane, Chichester,
West Sussex PO19 1UD, England

National   01243 779777
International (+44) 1243 779777
e-mail (for orders and customer service enquiries): cs-books@Wiley.co.uk
Visit our Home Page on http://www.wiley.co.uk
or http://www.wiley.com

Chapter 16 © 1997 Her Majesty's Inspectorate of Pollution
Chapter 27 © 1997 Anglo American Corporation of South Africa Limited

All Rights Reserved. No part of this book may be reproduced, stored in a retrieval system, or transmitted, in any form or by any means, electronic, mechanical, photocopying, recording, scanning or otherwise, except under the terms of the Copyright, Designs and Patents Act 1988 or under the terms of a licence issued by the Copyright Licensing Agency, 90 Tottenham Court Road, London WIP 9HE, UK, without the permission in writing of the Publisher.

Other Wiley Editorial Offices

John Wiley & Sons, Inc., 605 Third Avenue,
New York, NY 10158-0012, USA

WILEY-VCH Verlag GmbH,
Pappelalle 3, D-69469 Weinheim, Germany

Jacaranda Wiley Ltd, 33 Park Road, Milton,
Queensland 4064, Australia

John Wiley & Sons (Asia) Pte Ltd, 2 Clementi Loop #02-01,
Jin Xing Distripark, Singapore 129809

John Wiley & Sons (Canada) Ltd, 22 Worcester Road,
Rexdale, Ontario M9W 1LI, Canada

Library of Congress Cataloging-in-Publication Data

Environmental management systems and cleaner production / edited by
  Ruth Hillary.
    p.   cm.
  Includes bibliographical references and index.
  ISBN 0-471-96662-2
    1. Production management—Environmental aspects.   2. Manufactures—
Environmental aspects.   I. Hillary, Ruth.
  TS155.7.E58   1996
  658.4'08—dc20   96-36110
                                                                               CIP

British Library Cataloguing in Publication Data

A catalogue record for this book is available from the British Library

ISBN 0-471-96662-2

Typeset in 10/12pt Ehrhardt from the author's disks by Dorwyn Ltd, Rowlands Castle, Hampshire
Printed and bound by Antony Rowe Ltd, Eastbourne

This book is printed on acid-free paper responsibly manufactured from sustainable forestation, for which at least two trees are planted for each one used for paper production.

To John Worontschak and our lovely son Maximilian.

Contents

Acknowledgements	xvii
Glossary	xix
About the contributors	xxiii
Foreword	xxix
Jacqueline Aloisi de Larderel, Director, UNEP Industry and Environment	
Introduction	1
Ruth Hillary	
The approach of this book	2
Expert seminar	3
Recommendations	5
Contributors	6
The book's structure	7

SECTION I THE INTERNATIONAL DIMENSION

1 Introduction	11
Ruth Hillary	
2 Drivers for international integrated environmental management	15
John Wolfe	
Introduction	15
Proliferation of systems	16
International standards	17
Cleaner production	18
Key drivers	18
Practical, useful and usable	22
The conformity assessment question	23
Environmental standards and trade	24
Meeting the challenge	25
Conclusion	26
Reference	26

3   An insight into the development and implementation of the
    international environmental management system ISO 14001   27
    Oswald A. Dodds
        Introduction   27
        Environmental management—what is it and why is it important? 29
        ISO TC 207, Sub-Committee 1—Environmental Management Sytems   30
        The specification—content and approach of ISO 14001   31
        The guidelines—content and approach   34
        The ISO standards—their intended users?   34
        Note   35

4   International voluntary standards—the potential for trade
    barriers   37
    Kerstin Pfliegner
        Introduction   37
        Potential positive effects on trade   38
        Potential barriers to trade   39
        Strategies to avoid trade barriers   43
        Conclusion   46
        Notes   46
        References   47

5   The possibility of cleaner production worldwide   49
    Sybren de Hoo
        Introduction   49
        What is cleaner production and why invest in it?   50
        Cleaner production, sustainable development and the relevance for
        developing countries   52
        What are the elements of an overall strategy to establish a cleaner
        production policy?   52
        What are the main elements in the toolbox for cleaner production
        policies?   54
        Conclusions   56

SECTION II   NATIONAL PERSPECTIVES

6   Introduction   61
    Ruth Hillary

7   Management systems: getting lean, getting green in the USA   65
    John Atcheson
        Introduction   65
        What is lean and clean management?   66
        Elements of lean and clean   67
        The implications of lean and clean for public policy   69

	Beyond command and control	71
	Policies encouraging lean and clean management	72
	Conclusion	74
	Notes	74
	References	74

8 Swedish national environmental policy and environmental management systems in industry — 77
Mikael Backman

Background	77
Basic principles and the trend of environmental policy	78
The development of an industrial environmental policy	80
Environmental management tools placed in a system	82
Environmental management systems quality certification	83
Knowledge of various environmental management standards	84
Advice for small and medium-sized companies	86
Conclusions	87
Note	88
References	89

9 Environmental management initiatives in China to promote cleaner production — 91
Ya-Hui Zhuang

Introduction	91
Initiatives in cleaner production	92
Evaluation of the environmental management system to promote cleaner production	92
Economic tools in the promotion of cleaner production	93
Priority setting for cleaner production options	97
Ecological planning as an effective tool for township enterprise reform	98
Conclusions	100
References	100

10 Hong Kong's experience with cleaner processes — 101
Edwin Kon-hung Lui

Geographical and population data of Hong Kong	101
Hong Kong's industry and cleaner processes in general	101
Chemical waste management in Hong Kong	102
Other measures for cleaner processes in Hong Kong	105
Conclusion	105
Acknowledgements	106
References	106

11 Linking cleaner production and ISO 14001 environmental management systems in New Zealand — 107
Marje Russ
- New Zealand's business environment — 107
- New Zealand's regulatory environment — 108
- Environmental management systems in New Zealand — 109
- Cleaner production in New Zealand — 109
- Links between environmental management systems and cleaner production — 110
- Links in practice — 115
- Conclusion — 115
- Notes — 116
- References — 116

12 Cleaner production and environmental management systems in Australia — 117
Brian O'Neill
- Cleaner production initiatives in Australia — 117
- Awareness raising initiatives — 118
- Demonstration projects — 118
- Environmental management systems in Australia — 119
- Environmental management systems, environment protection and cleaner production in Australia — 121
- Environmental management systems' effectiveness and objectives — 121
- Conclusion — 122

SECTION III REGULATION OR SELF-REGULATION?

13 Introduction — 125
Ruth Hillary

14 EU environmental policy, voluntary mechanisms and the Eco-management and Audit Scheme — 129
Ruth Hillary
- Introduction — 129
- EU environmental policy — 130
- Normative legislation — 130
- Towards sustainability — 131
- Market-based tools — 132
- Eco-management and Audit Scheme — 132
- CEN mandate — 133
- Pilot project company experience — 135
- EVABAT — 140
- Conclusion — 141

	Notes	142
	References	142
15	**Clean production and the post-command-and-control paradigm**	143
	David Rejeski	
	Introduction	143
	Learning before doing	146
	Environmental management across the value chain	147
	Customer-driven production	149
	Whither public policy?	150
	Notes	153
	References	154
16	**The role of regulatory systems in requiring cleaner processes and relationships with voluntary systems**	157
	Allan G. Duncan	
	Introduction	157
	The regulatory system: integrated pollution control	159
	BATNEEC	161
	A force for cleaner production?	162
	Innovation and technology forcing	163
	Conclusion	164
17	**BS 7750 and certification—the UK experience**	165
	Christopher Sheldon	
	Introduction	165
	When is an audit not an audit?	168
	The European context	168
	Certification: the working proof	170
	Changing management perspectives	171
18	**Environmental management system certification—an assessor's view**	173
	Jeff Dowson	
	Introduction	173
	Environmental management system assessment	174
	Effects evaluation	174
	The management system	177
	Policy integration	178
	Register of legislation	178
	Objectives and targets	179
	Internal audits	179
	Integration of management systems	180
	Conclusion	181

19	**Certification and harmonization of environmental management systems**	183
	Roger Brockway	
	Introduction	183
	Environmental management systems	183
	Accredited certification	184
	ISO 14001/ISO 9000	187
	Progess in UKAS accreditation	187
	Harmonization of accreditation worldwide	187
	Conclusion	188

SECTION IV   EUROPEAN INDUSTRIAL EXPERIENCE

20	**Introduction**	191
	Ruth Hillary	
21	**Potential for improving environmental performance through implementation of integrated management systems**	195
	Johan Thoresen	
	Limitations of technology as a sole solution to environmental problems	195
	Integration of environmental issues into systems for strategic and operational management	197
	Integrated management and life cycle principles demand systems thinking in two directions	200
	Prevention rather than process improvements and end-of-pipe solutions	201
	Conclusions	203
	References	204
22	**The integrated approach: the Chemical Industries Association's Responsible Care**	205
	Dr Stuart Aaron	
	Responsible care aims	205
	Integratable standards for health, safety and the environment	206
	Sector guidance	208
	One-stop auditing	210
	Future demands on the standard makers	210
	Conclusion	211
23	**Risk-management approach to environmental management—UK case studies**	213
	Dr Paul Pritchard	
	Introduction	213
	Environmental risk-based management	214
	Risk transfer and the insurance industry	215

	Risk-based tools	216
	Environmental audits	216
	Calculation of risk profiles	217
	Conclusion	218
	Reference	218
24	**Experience of environmental management in the Danish fish-processing industry** *Eskild Holm Nielsen*	219
	Cleaner production in the fish-processing industry	219
	Design of the case study	220
	The environmental effect of cleaner technologies in the herring industry	221
	Discussion	224
	References	227
25	**Cleaner production through environmental management of process innovations** *Nils Thorsen*	229
	Introduction	229
	Fundamental changes in the general organization of environmental protection	229
	The Achilles' heel of environmental management systems	230
	Meeting global environmental challenges	231
	Is cleaner production competitive?	232
	Examples from business practice: managing process innovations towards cleaner production	233
	Conclusions	235

SECTION V INDUSTRIAL EXPERIENCE FROM EMERGING AND TRANSITION ECONOMIES

26	**Introduction** *Ruth Hillary*	239
27	**Environmental management systems for cleaner operations in South African mines** *Harold Nicholls*	243
	Introduction	243
	Binding commitments to protect the environment	244
	Supplementary documents to assist in preparation and implementation of an EMPR	247
	The implementation of an environmental management system	248
	Capacity building and environmental awareness	250
	Conclusion	256

28	**Environmental management initiatives in the Brazilian oil industry**	259
	Sergio Pinto Amaral	
	Introduction	259
	Brazilian and international laws, regulations and standards on environmental management and auditing	260
	The environmental management system in place in Petrobras	261
	Petrobras's environmental initiatives relating to cleaner products and processes	262
	Environmental promotion, awareness raising and training	264
	Conclusions	264
	References	265
29	**The greening of Lithuanian industry: past and present**	267
	Leonardas Rinkevicius	
	Introduction	267
	Profile of Lithuanian industry and environment: past and present	268
	The research approach and some findings	270
	Incentives and constraints for waste and pollution minimization in Lithuanian industry	274
	Conclusions	279
	References	280
30	**Implications of Czech cleaner production case studies for environmental management systems**	283
	Vladimír Dobeš	
	Introduction	283
	Background information	284
	Cleaner production case studies	284
	Implications for environmental management systems	287
	Conclusion	292
	References	292
31	**The combined introduction of environmental management systems and cleaner production in industry**	293
	Gulbrand Wangen	
	Introduction	293
	Today's trends	294
	Environmental management systems (EMS)	295
	Cleaner production (CP)	295
	Situation in Norway	296
	Situation in Hungary	299
	QEMS in Hungary	300
	Conclusions	303

SECTION VI PRACTICAL CASE STUDIES FROM SMALLER COMPANIES

32 Introduction 307
Ruth Hillary
   Reference 309

33 Stimulating environmental action in small to medium-sized enterprises 311
Michael Smith
   Introduction 311
   The importance of the SME sector 311
   Partnerships to assist SMEs 313
   The Groundwork approach 313
   Barriers to improving SME environmental performance 314
   Groundwork's experience in assisting SMEs in improving environmental performance 314
   Case studies: the facts and figures of success 317
   Conclusion 318

34 EMAS adoption by an SME in the chemical sector 319
Vittorio Biondi and Marco Frey
   Introduction 319
   Lati: company description 320
   Quality and environmental management 321
   Initial environmental review 323
   Environmental policy and programmes 323
   EMS implementation 324
   External communication 326
   Conclusions 327
   Note 327

35 Achieving improvements in production through environmental management systems 329
Chris Burleigh
   Introduction 329
   Company profile 330
   Training 330
   Towards certification 330
   Potential incident tests the EMS 331
   Environmental initiatives 335
   Conclusion 336

36	**The practical implementation of BS 7750, EMAS and ISO 14001 within a medium-sized manufacturing site**	337
	Ken Jordan	
	Introduction to Akzo Nobel Chemicals' Gillingham site	337
	BS 7750 environmental management system	338
	Initial review	340
	Environmental policy	340
	Training	341
	Management responsibility	341
	Legislative register	342
	Register of significant effects	342
	Management programme	343
	System integration and auditing	343
	EMAS registration	344
	The enviromental statement	344
	The benefits of EMAS registration	345
Index		347

Acknowledgements

My special thanks go to the United Nations Environment Programme Industry and Environment Programme Activity Centre (UNEP IE/PAC) Cleaner Production Programme, in particular the Working Group on Policies, Strategies and Instruments and its Secretary Mikael Backman, who provided the framework and moral support for the Expert Seminar on "Environmental Management Systems in the Promotion of Cleaner Processes and Products", and to Sep Baghi, Shadi Khoroushi, Anna Burns and all who provided organizational and administrative support during the hectic weeks of preparation for the seminar. My special thanks also to all the contributors to this volume for their commitment to the project and their ability to meet my very tight deadlines within their own very busy schedules, and to the team at John Wiley for their advice and patience. And a final sincere thanks to all those at the Centre for Environmental Technology, Imperial College, who have allowed me the freedom to explore and expand my ideas and projects.

Glossary

ABNT	Brazilian Association of Technical Norms
ALADI	Latin American Integration Association
ANSI	American National Standards Institute
ARF	Applied Research Fund
BAT	The principle of best available technology
BATNEEC	Best available technology not entailing excessive cost
BCSD	Business Council for Sustainable Development
BOD	Biochemical oxygen demand
BOT	Build operate and transfer
BPEO	Best practical environmental option
BSI	British Standards Institution
CD	Committee draft
CEE	Central and Eastern European Countries
CEFIC	Conseil Européen de l'industrie Chimique
CEMC	Czech Environment Management Centre
CEN	Comité Européen de Normalisation (The European Committee for Standardization
CEO	Chief executive officer
CIA	Chemical Industries Association
CIGNs	Chief Inspector's Guidance Notes
CII	Confederation of Indian Industries
CIMAH	Control of industrial major accident hazards
CIPI	Interministerial Committee for the Co-ordination of Industrial Policy
CIS	Commonwealth of Independent States
COD	Chemical oxygen demand
COSHH	Control of Substances Hazardous to Health
CP	Cleaner production programmes
CPC	Czech Cleaner Production Centre
CSD	Commission for Sustainable Development
CSFI	Centre for the Study of Financial Innovation
CWTC	Chemical waste treatment centre

DIN	Deutsches Institut für Normung eV
DIS	Draft international standard
DNV	Det Norske Veritas
DOP	Dioctyl phtalate
EAC	European Accreditation of Certification
EMA	Environmental management and auditing
EMAS	Eco-management and Audit Scheme
EMC	European Marine Contractors Ltd
EMPR	Environmental Management Programme Report
EMS	Environmental management system
EPA	Environment Protection Agency
EPA'90	Environmental Protection Act of 1990
EQS	Environmental quality standards
ESID	Ecologically sustainable industrial development
EU	European Union
EVABAT	Economically viable application of best available technology
FMEA	Failure modes effects analysis
GANA	Supporting Group on Environmental Standardization
GATT	General Agreement on Tariffs and Trade
GDP	Gross domestic product
GEMI	Global Environmental Management Initative
GNP	Gross National Product
HAZOP	Hazard and Operability Study
HDT	Hydrode sulphurization
HMIP	Her Majesty's Inspectorate of Pollution
HSE	Health and Safety Executive
IAF	International Accreditation forum
IBAMA	Brazilian Institute for the Environment and Renewable Natural Resources
IC	Internal control
ICC	International Chamber of Commerce
ICCET	Imperial College Centre for Environmental Technology
ICLAB	Irish Certification and Laboratory Accreditation Scheme
ICPIC	International Cleaner Production Information Clearinghouse
IEC	International Electrotechnical Commission
IE/PAC	Industry and Environment Programme Activity Centre
IMPEL	Information Manual and Programme of Environmental Legislation
IMS	Integrated management system
IPC	Integrated pollution control
IPPC	Integrated pollution prevention and control
ISO	International Organization for Standardization
ISONET	International Organization for Standardization's information network

JAS-ANZ	Joint Accreditation Scheme Australia New Zealand
JIT	Just-in-time
LCA	Life cycle assessment
LDC	Lesser developed countries
MACT	Maximum achievable control technology
NAFTA	North American Free Trade Agreement
NGOs	Non-governmental organizations
NIC	Newly industrialized countries
NIF	Norwegian Society for Chartered Engineers
NNI	Nederlands Normalisatie Instituut
NPCA	Norwegian Pollution Control Authority
NSCE	Norwegian Society of Chartered Engineers
NTS	Norwegian Technical Standards Institute
NZCIC	New Zealand Chemical Industry Council
OECD	Organization for Economic Co-operation and Development
OH&S	Occupational health and safety
OSPAR	Joint Oslo and Paris Commissions
PER	Preparatory environmental review
PERI	Public environmental reporting initiative
PIMS	Public Involvement Management System
QA	Quality assurance
QEMS	Quality and environmental management system
QMI	Quality Management Institute
QSAR	Quality System Assessment Recognition
RCRA	The Resource Conservation and Recovery Act
SA	Standards Australia
SAGE	Strategic Advisory Group on the Environment
SAQAS	Standards Australia Quality Assurance Services
SMEs	Small and medium-sized enterprises
SRP	Special restructuring programmes
SUSEMA	Superintendency for the Environment, Quality and Industrial Safety
TBT	Technical Barriers on Trade Agreement
TIF	Technology Innovation Fund
TQEM	Total quality environmental management
TQM	Total quality management
UKAS	United Kingdom Accreditation Service
UN	United Nations
UNCTAD	United Nations Conference on Trade and Development
UNDP	United Nations Development Programme
UNEP	United Nations Environment Programme
UNIDO	United Nations Industrial Development Organization
US EPA	United States Environmental Protection Agency

US FDA	United States Food and Drug Administration
VOC	Volatile organic compounds
WCED	World Commission on Environment and Development
WCM	World class manufacturing
WEC	World Environment Centre
WICE	World Industry Council for the Environment
WTO	World Trade Organization

About the Contributors

Stuart Aaron is Manager of the Chemical Industries Association's Responsible Care programme. He has a PhD in chemical engineering from Imperial College, London. He has worked for Shell Research Laboratories, in Amsterdam on oil desulphurisation and polypropylene process development in the manufacturing organisation. He worked for 11 years in operations management at the Shell Chemicals Stanlow Petrochemicals complex and latterly in Strategic Planning Division of Shell International Chemicals.

Sergio P. Amaral is a senior environmental engineer working for Petrobras – the Brazilian State-owned Oil Company. He leads teams that implement environmental management systems and auditing programmes in Petrobras. He also works for the Brazilian Association of Technical Norms (ABNT), as one of the Brazilian experts who attends the meetings on the ISO 14001 series of Environmental Standards.

John Atcheson is director of the Office of Policy Analysis within the USA Department of Energy's Office of Energy Efficiency and Renewable Energy. He helped set up The Office of Pollution Prevention at the Environment Protection Agency and was instrumental in developing a number of innovative alternatives to command and control regulations, including some of the earliest applications of environmental management systems. He has nearly 20 years of experience in the environmental field.

Mikael Backman is one of the founders of the International Institute for Industrial Environmental Economics at Lund University, Sweden. He is Head Researcher at the Institute, with a concentration of research on policies, strategies and instruments to promote preventative approaches to environmental problems in society. He is also responsible for the section on Environmental Management in the international Master's Program on Environmental Management and Policy offered by the Institute.

Vittorio Biondi is a researcher at the Institute of Energy and Environmental Economics (IEFE), Bocconi University, Milan. In 1994/5 he worked on a pilot project, co-financed by the European Commission DG XI, on the diffusion of EMAS in the Italian chemical industry. He is also an external consultant to DG XI working on the co-ordination of EMAS pilot projects sponsored by the European Commission. He is a lecturer on environment, health and safety management at SDA Bocconi and on several post graduate courses.

Roger Brockway is Head of the National Accreditation of Certification Bodies (NACB) at the United Kingdom Accreditation Service (UKAS) responsible for the accreditation of certification bodies for BS 7750 and environmental verifiers for Eco-management and Audit scheme (EMAS). He is Chairman of the European Accreditation Working Group on environmental matters and is a member of the UK Government's Advisory Group on EMAS. Formally, he was the head of the BSI Legal Department and then the Quality Policy Unit and was seconded to the European Commission's Directorate-General for Industry and the Internal Market (DGIII) where he was responsible for developing the European Organisation for Testing and Certification (EOTC).

Chris Burleigh is the senior partner of Target Environmental Systems. He was a quality manager for seven years before becoming involved with the environmental management systems BS 7750 and EMAS. He has successfully implemented environmental management systems in many companies including several blue chip companies.

Sybren de Hoo is a senior advisor to the United Nations Environment Programme (UNEP) at the Industry and Environment Office in Paris. He has been in charge of many cleaner production programmes in Europe and has served as a deputy director of the Dutch Organisation of Technology Assessment (Rathenau Institute). Over the last five years he has worked as an advisor for the World Bank and UNEP in cleaner production in China.

Vladimir Dobeš is Director of the Czech Cleaner Production Centre and has been a UNIDO national consultant for cleaner production in the Czech Republic. He gained his first experience with cleaner production during his participation in European Postgraduate Course in Environmental Management in the Netherlands and has been leading activities on promotion of cleaner production in the Czech Republic since 1994.

Oswald A. Dodds, MBE, is Chairman of ISO TC 207 Sub Committee 1 on Environmental Management Systems, he was also the Chair of the BSI Technical Committee responsible for the production of BS 7750 1994. He is also Director of Contract Services for Northampton Borough Council in the UK.

Jeff Dowson is Environmental Management Systems Business Manager for SGS Yarsley ICS Limited. He is a Registered Lead Auditor and a qualified industrial chemist.

Allan G. Duncan is the former Chief Executive and Chief Inspector of Her Majesty's Inspectorate of Pollution (HMIP) and the Head of Function for Radioactive Substances Regulation of the Environment Agency which supersedes HMIP. He is a

ABOUT THE CONTRIBUTORS

qualified chemist and a member of the Environmental Accreditation Panel of the UK's National Accreditation Council for Certifying Bodies which judges applications for accreditation of certifying bodies for BS 7750 and ISO 14001 and of verifiers for EMAS.

Marco Frey is a research fellow in management at Tor Vergata University, Rome, and vice director of the Environmental Division at the Institute of Energy and Environmental Economics (IEFE), Bocconi University, Milan. In 1994/5 he was co-ordinator of a pilot project, co-financed by the European Commission DG XI, on the diffusion of EMAS in the Italian chemical industry. He is also an external consultant to DG XI working on the scientific co-ordination of EMAS pilot projects sponsored by the European Commission. He is a lecturer on environment and safety management at SDA Bocconi and on several post graduate courses.

Ruth Hillary is a leading researcher at Imperial College's Centre for Environmental Technology where she undertakes EU and UK funded research into environmental management systems and small and medium sized enterprises, in particular as Project Manager of a European Commission's DG XI EMAS pilot project. She is the UK National Coordinator for the European Commission's DG XXIII Euromanagement–Environment pilot action. She is founder of the Network for Environmental Management and Auditing (NEMA) and a member of the UK Government's Advisory Group on EMAS. She is the author of *The Eco-management and Audit Scheme: A Practical Guide* and the series editor for the Business and the Environment Practitioners Series. She is the editor of *Environmental Management Systems and Cleaner Production* and can be contacted by e-mail: r.hillary@ic.ac.uk.

Ken Jordan is Technical Manager of Akzo Nobel Chemical, Gillingham, UK. He acted as Project leader for the implementation of the Gillingham site's environmental management system and its subsequent accredited certification to BS 7750 in March 1994 and the draft ISO 14001 in March 1996 and EMAS registration in August 1995.

Edwin K.H. Lui is an Environmental Protection Officer of the Hong Kong Environmental Protection Department.

Harold Nicholls has been an environmental civil engineer in the Anglo American Corporation for seven years. His experience has been mainly in mining operations and includes environmental management systems, environmental impact assessments, water management systems and mine closures. His previous experience was in sewage purification with the City Council of Johannesburg where he worked for 29 years.

Eskild Holm Nielsen is a researcher at Aalborg University within the field of Environmental Planning and Management. His research focuses on the interplay between environmental regulation and the impact on companies' innovation and environmental performance. He has written many articles on environmental management and auditing as well as market-based environmental regulation.

Brian O'Neill is Assistant Director, International and Stakeholder Management Section of Environment Forests Task Force, Environment Australia. He is an environmental scientist with a background in industry, education and government. He has been a member of Standards Australia Committees providing input to the ISO 14000 series and an Australian delegate to ISO TC 207 meeting in 1994/5. His work has involved him in Australian developments in environmental management system accreditation and certification, environmental auditor certification and registration as well as providing advice on environmental management systems to the Australian government.

Kerstin Pfliegner is a consultant for the Private Sector Development Programme of the United Nations Development Programme. She is the UNDP representative in the ISO technical committee developing ISO 14000. She studied Business Management and Environmental Economics and has been working for private sector companies and as a researcher and lecturer at Universities, prior to joining the UN.

Paul Pritchard is Environmental Adviser to Royal and Sun Alliance Insurance Group. Following completion of a PhD in environmental chemistry he worked for the UK Department of Trade and Industry. His research interest focuses on the application of risk techniques to environmental management.

David Rejeski is employed at the White House Office of Science and Technology Policy (OSTP) on a variety of issues including: environmental research policy, environmental education, and the implementation of the National Environmental Technology Strategy. Before joining OSTP, he was head of the Future Studies Unit in the Office of Policy, Planning and Evaluation at the US Environmental Protection Agency.

Leonardas Rinkevicius is with the Pollution Prevention Centre established at Kaunas University of Technology in co-operation with the World Environmental Centre (New York) and lectures at the university's Department of Public Administration as well as assisting in curriculum development in environment, technology and society studies. He works to promote pollution prevention and environmental management in Lithuanian industry and has worked as a consultant to the World Bank, Lithuanian Ministry of Environmental Protection, EU Phare programme and Harvard Institute of International Development.

Marje Russ is the General Manager, Environmental Certification Services for Telarc New Zealand. She is an environmental planner and scientist with extensive environmental management experience in regulatory bodies, private consulting and certification services in New Zealand, Australia and the UK. She has been closely involved in the development and implementation of ISO 14000 series standards and providing related environmental management training in New Zealand.

ABOUT THE CONTRIBUTORS

Christopher Sheldon is an environmental writer and trainer and has been the Senior Environmental Policy Advisor to the British Standards Institution for the last six years. He has been leading teams involved in the creation, implementation and certification of BS 7750 since its inception in 1990.

Michael E. Smith has been the Director of Environmental Management Services at Groundwork, Blackburn, since 1991. He joined Groundwork in 1986 focusing on landscape and environmental management. He is an environmental auditor under the Environmental Auditors Registration scheme and an advisor for the Environmental Technology Best Practice Programme and a BSI Technical Committee member.

Johan Thoresen is a senior researcher at the Oestfold Research Foundation in Norway where he has been involved in the development of a company's integrated management systems and undertaken cleaner production projects in both Norway and Sweden. He has an MSc in environmental management and is currently undertaking a PhD in environmental management and integrated management.

Nils Thorsen is Senior Manager at Ernst & Young Environmental Services. He worked for 14 years in public environmental regulation in the Danish Ministry of Environment and the Copenhagen municipality and 3 years as an Environmental Auditor for the large international insulin and enzyme manufacturer Novo Nordisk. He has an MSc in Environmental Planning from the Institute for Environment, Technology and Society at the University of Roskilde.

Gulbrand Wangen is a senior consultant at Det Norske Veritas (DNV) Industry in Norway.

John Wolfe is the former director of the Canadian Standards Association (CSA) Standards Division where he managed the activities of CSA's 9000 volunteers and the ISO secretariat for TC 176 on Quality Management and was the International Secretary of ISO TC 207 on Environmental Management. He is now a senior consultant managing the start-up of ICF Kaiser International Canadian office.

Ya-hui Zhuang is a professor of environmental science at the Research Centre for Eco-environmental Sciences, Chinese Academy of Sciences, and has been the principal investigator of several national research projects. He supervised the demonstration of process-water conservation, waste-water recycling and value-addition during waste recovery at the Yanshan Petrochemical Incorporation. He has taught on environmental management and air pollution control courses at the Asian Institute of Technology, Bangkok and is now teaching the course on environmental engineering and management at the Beijing Chemical Engineering Institute. He is the co-author of the SCOPE/CHINA monograph entitled *Element Cycling in Complex Ecosystems*.

Foreword

BY
JACQUELINE ALOISI DE LARDEREL
Director of UNEP Industry and Environment

Increasing the efficiency of the use of raw materials by a factor of five to ten within the next 30 to 40 years, is the challenge the world faces if the goal of sustainable development set by the Earth Summit in 1992 is to be achieved. This will certainly require all stakeholders to break away from "business as usual".

For government this will mean setting up regulatory frameworks, economic instruments and institutional arrangements to encourage industry and consumers to adopt sustainable production and consumption patterns. For industry this will mean developing cleaner production processes, products and services.

Cleaner production is a proactive approach which involves preventing the pollution of air, water and land, reducing waste at source, minimizing risks to the population and the environment, and minimizing the use of raw materials, including energy and water. Besides reduced environmental impacts, cleaner production leads to monetary savings because fewer raw materials are used and there is a reduced need for pollution control. Cleaner production through ecoefficiency is the way to reconcile environmental protection and economic development.

Cleaner production does not only require industry to introduce new technologies; it also requires a manager to integrate environmental issues in day-to-day business management from plant operation to purchasing, from product design to waste disposal, for example.

Industry managers are now recognizing that to stay in business, they have to define environmental targets, develop environmental management systems, adopt voluntary codes of practice and monitor their results.

For a number of years, UNEP has been promoting the cleaner production approach. It has provided a platform for the exchange of information and experience and facilitated the involvement of an international network of national organisations in cleaner production.

This is why UNEP welcomes the publication of this book: *Environmental Management Systems and Cleaner Production*, edited by Ruth Hillary at the Centre for

Environmental Technology, Imperial College of Science, Technology and Medicine, UK. This book will contribute to the educational process needed to change attitudes. Readers will certainly gain valuable information which will lead them to take action. And in the quest for a sustainable world, every action counts.

Introduction

RUTH HILLARY

Environmental management systems and cleaner production are important subjects and worthy of the detailed discussion they receive in this book. The two fields have emerged against a backdrop of increasingly stringent environmental legislation, growing public awareness, globalization of trade and harmonization of standards, leading businesses to identify ways to mitigate and manage the environmental impacts associated with their production activities. Normative environmental regulation has tended to push environmental protection towards end-of-pipe solutions rather than cleaner production. Furthermore, such regulation shifts the responsibility for environmental protection on to the regulator rather than the regulated, whereas cleaner production and environmental management systems require management to take more responsibility for achieving environmental goals.

Traditional regulation has not delivered the environmental improvements hoped for by governments, and businesses argue that the solutions which they adopt to meet environmental standards are not the most cost-effective way of achieving environmental objectives. Normative regulation is difficult and expensive to monitor effectively. Non-compliance is often attractive to enterprises. Pressure is building for deregulation, but alternative approaches to environmental protection, such as environmental management systems and cleaner production approaches, need to be fully tested before environmental regulation can be relaxed or removed.

Businesses have utilized management system approaches for health and safety and quality for many years. Environmental management systems have been developed more recently and consist of the organizational structure, responsibilities, practices, procedures, processes and resources which the enterprise uses to achieve its environmental policy. Formalized environmental management systems are documented voluntary systems. The first environmental management systems standard (BS 7750) appeared in 1992 in the UK, subsequently many other countries have developed standards and the International Organization for Standardisation (ISO) has also

---

Environmental Management Systems and Cleaner Production, edited by R. Hillary.
© 1997 John Wiley & Sons Ltd.

developed an environmental management systems standard, ISO 14001. Regional schemes also exist, for example the European Union's Eco-management and Audit Scheme (EMAS). Differing environmental regulations can be trade barriers, especially for developing countries, but it is necessary to identify to what extent the differing voluntary standards restrict trade.

Cleaner production, like environmental management systems, is a strategic approach to environmental protection. Cleaner production means designing industrial processes and products to prevent the pollution of media and to make efficient use of raw materials; therefore cleaner production approaches require enterprises to anticipate and mitigate the environmental impacts of their production processes. In general, production processes are retrofitted to control environmental impacts in response to legislation. Even normative legislation which gives preference to cleaner technology over end-of-pipe solutions is not normally heeded by industry.

Both environmental management systems and cleaner production approaches require businesses to be converted to their message: that it is cost-effective and beneficial for enterprises to manage their environmental performance proactively. Advocates of environmental management systems and cleaner production have to present persuasive arguments to convince the unconverted. And this is one of the features they have in common: policy makers and practitioners alike have to be convinced of the environmental benefits of environmental management systems and cleaner production before changing their approach to environmental protection.

Thus the purpose of this book is to introduce the reader to the new and dynamic fields of environmental management systems and cleaner production. It has been written for policy makers, industrial practitioners and researchers who wish to learn more about environmental management systems and their relationship with cleaner production. It aims to straddle the practical and the theoretical as well as the international, regional, national and industry perspectives. The book's contributors are its strength. They come from a diverse range of organizations and have vast experience which provides for a many layered approach to the topics discussed in this book.

THE APPROACH OF THIS BOOK

This book moves systematically from the international arena to selected national perspectives and then focuses on industrial case studies from developed and developing countries and from both small and large enterprises. The rationale behind this approach is to show how global activity on environmental management systems like ISO 14001 and cleaner production is linked to local actions, and how the international scene relates to national developments which in turn shape practical industrial experience. This volume also shows the links and relationships between system solutions and technological solutions to environmental impacts, illustrating the interplay that exists between the two.

The conceptual approach of the book and some of the issues it addresses are summarized in Figure 1.

INTRODUCTION

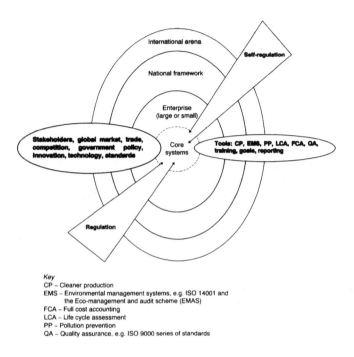

Figure 1 Approach and scope of issues in *Environmental Management Systems and Cleaner Production*

EXPERT SEMINAR

This book is an expansion on the output from the expert seminar on "Environmental Management Systems in the Promotion of Cleaner Processes and Products", organized within the framework of UNEP/IE/PAC Cleaner Production Programme and held at the Imperial College of Science, Technology and Medicine in September 1994. The book's chapters substantially enlarge on the themes of the seminar. Many chapters have been specifically written for this volume while others are updated and revised versions of seminar papers. All contributors to this volume are experts within the scope of the seminar.

An overall objective of the expert seminar, which has been carried forward and amplified in this book, was to focus on the evaluation of different national e.g. BS 7750, regional e.g. EMAS and international: ISO 14001 environmental management systems and cleaner production programmes and to identify their roles in delivering cleaner production in both processes and products. A key feature of the seminar was that it acted as a forum for discussion. To this end, four round-table sessions sought to summarize the main conclusions of the seminar presentations and expand on the issues raised to draw conclusions on the type of actions best able to promote cleaner

Expert seminar round table priority topics
- The role of environmental management systems in the promotion of cleaner technology in preference to end-of-pipe technology.
- The effect of employee involvement in implementation of environmental management systems.
- How can environmental management systems be related to regulation?
- Top management leadership versus employee involvement in environmental management systems.

Some additional topics explored in this book
- The role of enforcement in promoting cleaner technology and/or environmental management systems such as ISO 14001 and EMAS.
- Certification of environmental management systems on a company level.
- The use of self-regulatory tools such as environmental management systems to promote cleaner production.
- Technological verses management approaches to achieving improvements in environmental performance.
- The preparedness of an environmental management system to adapt to high uncertainty and/or high speed of change of company-external factors.
- Need for decision-support tools in environmental management.
- Environmental management systems and the needs of small and medium-sized enterprises.
- Innovation strategies and environmental policy instruments.
- Necessary balance between demands for management system structure and demand for organizational creativity and flexibility.
- Relationship of environmental quality objectives to environmental management systems.
- Performance evaluation and continuous environmental improvement as components of environmental management systems.
- Public reporting—is it an essential component of an environmental management system?
- Environmental specialists or managers—who should be running an environmental management system?
- Risk-management approach as a tool for achieving broader business acceptability.
- How to encourage companies to expand their quality management systems towards environmental management systems such as ISO 14001.
- Environmental management: changing paradigm or learning by doing?

Figure 2 Round-table discussion topics covered in *Environmental Management Systems and Cleaner Production*

INTRODUCTION 5

production via environmental management systems. The round tables focused on four priority topics (see Figure 2), these and many other topics were considered important and are explored by the contributors to this volume.

The seminar was organized with two additional aims in mind: first, to assess environmental management systems to determine if cleaner production is promoted rather than traditional pollution control technology; and second, to disseminate this knowledge widely and to a larger number of countries. This book is a major contribution to the effective dissemination of the knowledge amassed at the seminar.

RECOMMENDATIONS

At the final session of the expert seminar its participants adopted eight recommendations for consideration by policy makers, practitioners and researchers supporting the promotion of cleaner production and environmental management systems (see Figure 3). Many of the issues and assertions raised in the seminar's recommendations are explored in this book by the contributors, each providing a particular view on environmental management systems and cleaner production.

Recommendations adopted by the participants of the UNEP IE/PAC Cleaner Production Programme invitational expert seminar: *Environmental Management Systems in the Promotion of Cleaner Processes and Products* **held in London, 22–23 September 1994**

20 participants from 15 countries—Australia, Brazil, Canada, China, the Czech Republic, Denmark, France, Hong Kong, Italy, Ireland, Lithuania, Norway, Sweden, United Kingdom and the USA—working in international organizations, government ministries, industry, consultancies, research organizations and universities, took part in the expert seminar. At the final session, the following recommendations were adopted by the participants for consideration by policy makers, practitioners and researchers supporting the promotion of cleaner production and environmental management systems to improve environmental performance and stimulate sustainable development in enterprises.

- *Governments should develop policies which allow enterprises to build on the existing mandatory framework,* by aligning economic and environmental goals and stimulating voluntary undertakings.
- *Standards and other measures to promote environmental management systems need to respect international differences,* but all need to incorporate clear and credible continuous environmental improvement principles.

(continued)

(continued)

- *Governments should promote the uptake of voluntary environmental management systems*, because such systems allow the transfer of responsibility to enterprises which tend to set tougher and wider-reaching goals in voluntary situations.
- *Governments and international organizations should establish programmes to communicate the objectives of environmental management systems clearly.* These programmes should emphasize the positive benefits achievable by enterprises implementing environmental management systems, including the opportunities which such systems present to promote cleaner processes and products.
- *Lessons should be learned from cleaner production demonstration projects.* Messages about strategies for continuous environmental improvement, together with information on financial as well as environmental benefits, need to be part of demonstration projects and actively promoted to enterprises in relevant promotional material.
- *Training programmes should be supported which encourage the integration of environmental management systems into enterprises' core management systems,* foster the setting of environmental objectives and the practices of cleaner production, and maximize environmental improvements and the participation of employees in environmental programmes to achieve continual improvement.
- *UNEP should become a formal liaison member to the International Organization for Standardisation's (ISO) technical committee TC 207,* which is formulating standards in the environmental field.
- *Governments should encourage and enterprises should adopt environmental management systems which ensure credibility, accountability and comparability.* Credibility—data on performance should be compared to environmental performance; accountability—goals, targets and performance data should be developed through a formal process that includes stakeholders; comparability—reporting should enable realistic comparisons of environmental achievements between enterprises.

Figure 3 Expert seminar recommendations

CONTRIBUTORS

The contributors to this volume are its overriding strength. They come from 16 countries—Australia, Brazil, Canada, China, the Czech Republic, Denmark, France, Hong Kong, Italy, Lithuania, New Zealand, Norway, South Africa, Sweden, the UK and the USA—and work in international organizations, government bodies and

INTRODUCTION 7

agencies, research institutions, consultancies and industry. They have detailed experience and deep interest in the fields of environmental management systems and cleaner production. Some have overlapping experience which adds to the strength of this book by reinforcing ideas and arguments and by helping to define emerging concepts.

THE BOOK'S STRUCTURE

The book is divided into six sections, each of which explores a different dimension of environmental management systems and cleaner production in a number of self-contained chapters. The chapters can be read independently, although ultimately each is linked with its section's general theme.

The first section addresses the *international dimension* of environmental management systems and cleaner production. It identifies the drivers for the establishment of voluntary environmental management systems such as the international standard ISO 14001, its implications for business and the barriers which such systems can cause to trade. The section also discusses the type of government strategies and policies used to promote cleaner production in industry.

The second section considers selected *national perspectives* on environmental management systems and cleaner production, highlighting the polices, strategies and initiatives of the USA, Sweden, China, Hong Kong, New Zealand and Australia. Each country analysis draws together the government policies and practical industrial demonstration projects utilized to stimulate environmental performance improvements in enterprises.

The third section explores the relative merits of using either *regulation* or *self-regulation* mechanisms as a means to deliver cleaner production and improved environmental performance in industry. European Union and national environmental regulatory policies are critically reviewed and voluntary market-based initiatives such as the Eco-management and Audit Scheme and associated certification and accreditation systems are discussed to identify their role in environmental protection.

Sections IV, V and VI all address industrial experience, drawing on practical sector and company case studies to illustrate real implementation experience of environmental management systems, e.g. ISO 14001 and EMAS and cleaner production. The fourth section discusses *European industrial experience* from the oil, electricity generation, manufacturing, chemical, food processing and environmental consulting sectors. The fifth section focuses on *industrial experiences from emerging and transition economies* and the challenges that enterprises face on entering international markets which require increasingly stringent environmental standards. Smaller enterprises make up the vast proportion of the world businesses, therefore the final section explores *practical case studies from smaller companies* and how these businesses can be motivated to adopt environmental management systems and cleaner production techniques.

The editor provides an introduction to each section giving a brief overview of each contribution and establishing the salient and general themes and issues raised in the section. A range of acronyms exist in the fields of environmental management systems and cleaner production. Where they are introduced by the authors they are explained, and a glossary is provided on pages xix–xxii to assist the reader.

This book brings together wide-ranging analysis and practical experience on environmental management systems and cleaner production, two important approaches used to deliver environmental performance improvements in organizations. It contributes to the wider debate on the most appropriate mechanisms for delivering sustainable development. The book offers a broad range of views on the international, national and industrial approaches to environmental management systems and cleaner production and their implications for business. These approaches offer solutions to pollution and resource usage, but if two separate camps are developing, i.e. systems solutions versus technological solutions, the book shows the links and relationships between the two. In the following chapters, the contributors provide real insights into the fields of environmental management systems and cleaner production.

The views in this volume are not solely academic or theoretical: it also includes views based on detailed practical knowledge and case experience. One of its central aims will have been achieved if the reader understands the common goals of environmental management systems and cleaner production and the interrelationship between the international, regional and national levels which ultimately manifests itself in enterprises, both large and small, which are striving to improve their environmental performance and thus play a role in sustainable development.

Section I
The International Dimension

1

Introduction

RUTH HILLARY

The rapid and increasing globalization of trade and markets means that the internationalization of environmental management systems and cleaner production strategies is vitally important to businesses. It is therefore appropriate that the first section of *Environmental Management Systems and Cleaner Production* considers the international dimension. In four chapters, this section covers the development and drivers for international integrated environmental management (Chapter 2); the details of the international environmental management system (EMS) standard ISO 14001 (Chapter 3); the potential barriers to international trade from voluntary standards and amelioration mechanisms (Chapter 4) and government strategies and policies required to promote cleaner production in industry (Chapter 5). The reccurring theme of this section is how international activity to develop formalized international environmental management systems and cleaner production strategies and policies has arisen from the need for individual businesses to address their environmental performance and make the link between this performance and increased competitiveness.

In Chapter 2, John Wolf discusses the key drivers and chain of events which have resulted in the need for international integrated environmental management. He takes a panoramic view of his subject, identifying the five clusters of expectations which, he argues, a business must increasingly understand and perform to in order to remain competitive. He argues that these expectations have led to a proliferation of management systems and asserts that international standards required significant reorientation before they could play a role in strategic business management. Wolf characterizes systems standards as horizontal, cutting across organizational functions, and therefore strategic and technical standards as vertical and functionally compartmentalized; consequently he suggests that systems standards direct a business towards cleaner production and pollution prevention.

Wolf traces the development of international environmental management standards by the International Organization for Standardization (ISO). He analyses a suite

---

Environmental Management Systems and Cleaner Production, edited by R. Hillary.
© 1997 John Wiley & Sons Ltd.

of standards work by ISO's Technical Committee TC 207, asserting that for international standards to be instruments of business strategy, especially for small and medium-sized enterprises (SMEs), they must offer flexible guidance which enables innovation. He discusses the potential for standards to appear "exclusionary in spirit", both to SMEs and to trading partners. Finally, Wolf concludes that the success of TC 207's ambitious standards programme is dependent on industry using the standards, which in turn depends on industry developing a strategic and competitive interest in improving its environmental performance.

Oswald Dodds, in Chapter 3, expands on the systems thinking discussed by Wolf, focusing on the international EMS standard ISO 14001. Dodds argues that there is a business need for written management systems based on recognized "models". He asserts that standard-writing bodies, through their consensus-based standards-production procedure, have a major role to play in producing such models but highlights the weaknesses of this approach. He suggests that EMS standards production has been driven by the real need for many organizations to improve and demonstrate their environmental achievements and performance to stakeholders. Dodds then goes on to discuss the work of ISO's TC 207 Sub-Committee 1 (SC 1) on Environmental Management Systems, focusing on ISO 14001, its content and relationship with the European Union (EU) initiative, the Eco-Management and Audit Scheme (EMAS). Dodds concludes by discussing the needs of the intended users of the environmental management standards.

In Chapter 4, Kerstin Pfliegner discusses the potential barriers to trade from international voluntary standards such as ISO 14001. She argues that the intended users highlighted by Dodds are located in countries with widely differing levels of economic and regulatory development, but that standards development is weighted towards developed country interests to the detriment of developing countries. Based on the results of a United Nations Development Programme (UNDP) survey of experts, she suggests that the voluntary nature of ISO 14001 means that it could not be defined as an official trade barrier under the Technical Barriers on Trade Agreement (TBT) of the World Trade Organization (WTO). Pfliegner argues that there may be positive benefits to exporters in developing countries if ISO succeeds in harmonizing national and regional EMS standards, but she also suggests that the same exporters are more likely to experience trade barriers, citing five policy/enterprise obstacles in developing countries which act as barriers to trade. Pfliegner presents a range of strategies to avoid trade barriers in which, she argues, industrialized countries have an important role to play. Pfliegner concludes that only the careful application of ISO 14001 by WTO member countries will avoid trade barriers and achieve the objective of international standards setting.

In the final chapter of this section, Sybren de Hoo presents the strategies and policies which he argues have the possibility of delivering cleaner production worldwide. The strategies and policies discussed are the outcome of an Organization for Economic Co-operation and Development (OECD) and United Nations Environment Programme (UNEP) workshop and share many similarities with environmental management systems thinking. He defines cleaner production as an "anticipate and prevent" approach rather than a "react and treat" approach, and argues that the cleaner production approach combines the maximum effect for the environment with substantial economic savings for industry. He claims that investing in cleaner production is cheaper than relying on increasingly expensive end-of-pipe pollution control technologies, suggesting that cleaner production can minimize the need to make trade-offs between economic growth and protecting the environment and in doing so can contribute to sustainable development. This balance, he considers, makes cleaner production attractive to developing countries. De Hoo draws together the elements of a strategy for establishing a government's cleaner production policy, focusing on four groups of tools to promote and encourage industry to initiate cleaner production programmes. He concludes by suggesting that, to be successful, cleaner production requires a shared vision.

The four chapters of this section on the international dimension of environmental management systems and cleaner production share a number of common themes, including discussion on:

- the key drivers of and business need for environmental management systems and cleaner production;
- the link between stakeholder requirements and the need for businesses to demonstrate improvements in environmental performance;
- the positive benefits of the international harmonization of environmental management systems standards, coupled with the potential disadvantages to SMEs and developing countries' exporters;
- trade globalization and the creation of trade barriers;
- the requirements of the intended users of environmental management systems standards and cleaner production techniques;
- the reasons for the reliance on end-of-pipe pollution control technology;
- the role of standards and standard-making procedures including ISO 14001.

A binding theme throughout this section is the way in which environmental management systems standards and cleaner production adoption by businesses are viewed as constituting a proactive and strategic approach towards environment protection and how these approaches are fundamental to maintaining and increasing businesses' competitiveness. The section provides a panoramic view of international aspects of environmental management systems and cleaner production, highlighting the need for industry and governments to develop policies and strategies to address them.

These key issues form a central theme of the following section on national perspectives, which discusses selected country experiences with environmental management systems and cleaner production.

2
Drivers for international integrated environmental management

JOHN WOLFE

INTRODUCTION

It is a healthy sign that experts from different areas of management concern, such as quality and the environment, are motivated to coordinate their efforts. After all, a business really only wants one management system. To be competitive it has to be as efficient as possible, therefore integration is essential.

I would like to begin by taking a somewhat long view of my subject. In doing this it is hard to avoid certain clichés that characterize business today. We are all familiar with them:

- Business is becoming global.
- Global business is more competitive because there is more choice.
- With lots of choice, once distinctive products and services are treated like commodities.
- Commodities can only compete on price.
- To get out of the commodity spiral you must provide added value.
- Added value is dependent on differentiation.
- Differentiation is no longer dependent only on product or service attributes but also on performance and continual improvement.
- Performance no longer refers only to product or service performance, it also refers to organizational performance.

Environmental Management Systems and Cleaner Production, edited by R. Hillary.
© 1997 John Wiley & Sons Ltd.

To be competitive, then, you must have a well-run organization and not just a better mousetrap or mouse-catching service. In addition to the demands for continually improved product, service and organizational performance, the repertoire of performance issues is expanding. The environment is an important item in this repertoire.

To a large extent, the environment has become as prominent as it has because of the ever-broadening spectrum of stakeholders and expectations to which businesses must relate. It is no longer just the financial experts and shareholders who make demands at the annual meeting or call up the chief executive officer (CEO) or the investor relations director with importunate questions. There are now environmental groups, employees, neighbourhood and community organizations, unions, banks, insurers, pension fund managers and consumers demanding information and action. They all have power in their own way. They can all influence how successful a company will be. To be competitive a company must listen, understand and perform to expectations.

We can identify at least five clusters of expectations.

- *Economic:* A business cannot survive if it is not financially viable. Management must include in the decision-making mix not only a knowledge of costs and projected revenues but also an awareness of risks and liabilities.
- *Quality:* Customer and client expectations are increasing. Market access and competitiveness more and more mean providing assurance of quality.
- *Occupational health and safety:* Unions, employees and government regulators want assurance that the workplace is safe and presents no risks to health.
- *Environment:* An ever-broadening range of constituencies are keeping an eye on environmental performance and seeking corporate commitment to improve.
- *Social responsibility:* Management is increasingly being faced with issues and pressures of social responsibility and equity.

Standing back and looking at this line-up it is easy to see that some coordination is required. Although these demands all represent different areas of concern, they can be addressed in a similar way. The organization identifies the issues and expectations, establishes policies and objectives to meet them, manages their implementation, and monitors, documents and provides verifiable assurance of commitment, performance and improvement.

PROLIFERATION OF SYSTEMS

Unfortunately there has tended to be a proliferation of management systems to address these expectations. Quality, health and safety, finance and the environment are often segregated into silos of independent activity. This is a drag on efficiency, productivity and competitiveness and helps nobody but our competitors.

Professional bodies, regulators and organizations which develop and set standards have traditionally done little to avoid this "balkanization". They have protected turf and created areas of complex and arcane specialization. The institutes of chartered accountants have not coordinated their activities with the quality people, the quality people have not talked to the occupational health and safety (OH&S) people. The OH&S people sometimes talked about the environment but only because the "business and environment" people were not yet sufficiently organized. Hardly anybody talks much about social equity.

In this mix, the real challenge is to begin the coordinating activity. The International Organization for Standardization (ISO) environment and quality management systems standards have the opportunity to demonstrate by example how this integration can take place.

INTERNATIONAL STANDARDS

Fifteen years ago, when, for the most part, standards were still for engineers and had to do with product specifications, safety and component interface, international standards could not have played this role. But since the establishment of ISO Technical Committee 176 on Quality Management and Quality Assurance in 1979, there has been a significant reorientation. Standards are increasingly becoming a core element of strategic business management. As such standards must serve the need for efficient integration or be bypassed as impractical.

A clear sign that business is indeed viewing these standards strategically can be found in the new level of business engagement in the development process. At the inaugural meeting of the ISO Technical Committee on Environmental Management (TC 207) in June 1993, a meeting that was organized and held within the very short time frame of approximately three months, more delegates from more countries participated than at any other inaugural meeting of an international technical committee. Participation continues to grow weekly, with more and more senior-level people becoming involved.

This level of engagement is part of a more general trend. Stuart Wilkinson (1991) declared that:

> the power of standards has been discovered. We are now sending to the standards meetings our very best knights and heroes. Why is that?

In 1979 it would not have been possible to ask this question. Today it is rhetorical. A more useful question is: Why are systems standards strategic?

We can begin to answer this by characterizing systems standards as horizontal and technical standards as vertical. The horizontal standard cuts across the functions of an organization and is integrating, coordinating and holistic. It affects the strategic direction and overall performance of the business. A vertical standard compartmentalizes and stays within clearly defined organizational functions. It is seen as a technical not a business concern. This distinction is of considerable importance to both environmental standardization and management systems standardization.

Without wishing to force the point, I think that we can recognize a natural affinity between this kind of horizontal thinking and our changing relationship with the environment. Many analysts have remarked that the reason we are in our current environmental mess is that we neither saw nor recognized the importance of the ecosystems of which we are a part. In short, we were neither system thinkers nor horizontal thinkers: we were vertical thinkers. The engineer or designer was responsible for product specifications and not for the waste problem created or the resources needed to manufacture it. We tended to work and think and develop standards in silos of discrete activity and not in interconnected open systems.

CLEANER PRODUCTION

In the environmental area, systems thinking is leading us away from traditional end-of-pipe engineering and thinking towards prevention thinking. Cleaner production and pollution prevention require a rethinking of design and operational processes. Vertical, product specification thinking simply adds another technological device to the end of the line to reduce the parts per million of effluent to meet the new regulation. Systems thinking promotes the integration of business and other performance concerns. Cost, efficiency, productivity and environmental performance all become part of the same decision-making process. Another or a better end-of-pipe device, on its own, makes little impact; it is just another cost. Only through systems thinking can we achieve the degree of change needed to improve the environment significantly. It would be ironic if we got the connection between business systems and ecosystems right but forgot about the equally important integration of environmental management and other business management concerns.

It has taken a long time to wake up to the reality that environmental systems are a business systems issue, and that business, supported by appropriate public policy and standards, can be one of the greatest agents of change and improvement for the environment.

KEY DRIVERS

To understand how we got where we are it is useful to go back and look at the chain of events that led directly to international environmental management standards activity.

If any single event forced the connection between business and the environment on to the world's political and economic agenda it was the United Nations Conference on Human Environment held in Stockholm in 1972. This conference signalled the industrialized world's recognition that it finally had to face up to the environmental problems that uncontrolled industrial development was causing. Or as Richard Sandbrook, a participant in that first conference, put it, it was time to take responsibility for the "effluence of the affluent".

This conference resulted in the adoption of a global action plan for the environment and in the creation of the United Nations Environment Program (UNEP), which was charged with building environmental awareness and stewardship. An independent commission was also established following the conference, the World Commission on Environment and Development (WCED). It took upon itself the task of reassessing the environment in the context of development and providing a progress report. The commission, headed by Gro Brundtland, who later became the Prime Minister of Norway, published its landmark report, entitled *Our Common Future*, in 1987. It is in this report that we first find the term "sustainable development" and the call for industry to develop effective environmental management systems.

By the end of 1988 over 50 world leaders had publicly supported the report and were calling for a major international event to discuss and act on it. In 1989 the UN decided that there would be a United Nations Conference on Environment and Development (UNCED), also know as the Earth Summit. This conference was held in Rio de Janeiro in June 1992.

In the period leading up to the earth summit, ISO and the International Electrotechnical Commission (IEC) became directly involved. Maurice Strong, the Secretary General of UNCED and the driving force behind its organization, wanted to ensure that business was fully engaged in the process. He had asked Stephan Schmidheiny, a Swiss industrialist, to be his principal adviser on business issues. To perform this role, Schmidheiny established the Business Council for Sustainable Development (BCSD). One of the things this council did was to go to the two international standards organizations to see what they were doing in the area of environmental management and to encourage them to become more active.

This request from BCSD was very timely. It was consistent with work that was ongoing within the ISO/IEC President's Advisory Board on Technical Trends, the *ad hoc* group on Long Range Planning and the *ad hoc* group on Environmental Labelling. And so in August of 1991 the ISO and IEC formally established the Strategic Advisory Group on the Environment (SAGE) to study the situation and make recommendations.

With Frank Bosshardt of BCSD as chair, SAGE was asked to assess the need for standardization in the area of environmental management. Any work in the area had to ensure a common approach to management that would enhance both business and environmental performance and facilitate trade. The work of SAGE produced two major outcomes:

- a series of ISO/IEC recommendations on environmental management submitted to the UNCED preparatory conference in January 1992;
- in October 1992, a recommendation to the ISO/IEC to create a new ISO Technical committee to develop standards in the area of environmental management.

Both of these outcomes were effective:

- The call for better environmental management became a key element of the main documents that came out of UNCED: Agenda 21, the comprehensive policy

guidance document, and the Rio Declaration, a set of principles for achieving sustainable development.
- In January 1993, the ISO created a new technical committee, TC 207, to develop standards in the area of environmental management. Canada was the awarded the management of the secretariat for ISO TC 207 and the inaugural plenary session was held in Toronto in June 1993. The work had begun in earnest.

When the BCSD first approached the ISO and IEC it was not by accident. BCSD came to these organizations because they felt that the ISO/IEC system was effective for building international consensus and agreement. BCSD also recognized their work in the fields of water and soil quality as particularly valuable.

The new ISO TC 207 on Environmental Management has been given a more strategic mandate: to develop horizontal standards for environmental management tools and systems.

To fulfil its mandate TC 207's programme of work has been divided up among more than 20 committees. A great deal has been accomplished since its inception.

I believe that the committee reached an important watershed at its third plenary meeting, in June 1995. Over 500 people from nearly 50 countries, including 21 developing countries and 22 international liaison organizations, gathered in Oslo, Norway for the third plenary meeting of TC 207. This compares with some 320 delegates from 25 countries attending the second plenary in Brisbane, Australia in 1994. The plenary itself took place in two sessions, on 26 June and 1 July. The week in between was filled with over 75 scheduled meetings by TC 207 sub-committees and working groups.

The delegates to the meetings in Oslo came prepared to work hard. They understood the importance of maintaining TC 207's impressive rate of progress, without compromising the quality of the EMS standards it produces. The tone for the meetings was set in a speech by the Prime Minister of Norway, Gro Harlem Brundtland. In addressing a reception for TC 207 delegates, she said, "Your efforts must aim at real progress, not at petrifying an idea whose time has passed. Our common concern must be to constantly improve industry's environmental performance and industry must lead that way unless it wants to be led."

In the week of meetings that followed, the essential ingredients of hard work and compromise, leavened with humour, were used to good effect, resulting in strong progress towards TC 207's ambitious goal of producing international standards on environmental management within the next five years.

A major advance in the Oslo meetings was the promotion of six key EMS documents to Draft International Standard (DIS) status. This means that they are considered ready for a six-month final review and ballot before publication as ISO standards. Two of the most critical documents are:

- ISO DIS 14001 Environmental management systems—specification with guidance for use;
- ISO DIS 14004 Environmental management systems—general guidelines on environmental management principles, systems and supporting techniques.

The management systems elements that an organization must put in place to have its environmental management system (EMS) officially certified/registered are outlined in the ISO DIS 14001 specification document (in Canada, the Quality Management Institute (QMI), a division of CSA, and similar organizations will register a company's EMS to the requirements).

A much broader approach is taken in the ISO DIS 14004 guidance document, which provides useful background information, examples and practical tips. By providing both guidance and specification in environmental management systems, the ISO 14000 series of standards will meet a broad range of business needs—from general advice and guidance all the way to third-party registration. The core 14001 and 14004 standards were targeted for publication in mid-1996, and in fact was published in the autumn of 1996.

Three other important documents also went to DIS status, in the area of environmental auditing at the Oslo meeting:

- ISO DIS 14010: Guidelines for environmental auditing—general principles;
- ISO DIS 14011.1: Guidelines for environmental auditing—audit procedures—Part 1: auditing of environmental management systems;
- ISO DIS 14012: Guidelines for environmental auditing—qualification criteria for environmental auditors.

These documents, also now published, will help provide the necessary framework for fair, consistent environmental auditing. They will complement the ISO 14000 EMS standards and will make third-party registration possible. New work will begin in the related areas of environmental site assessments, initial review and the management of audit programmes.

Also moved to DIS status at Oslo was ISO DIS 14060: *Guide for the inclusion of environmental aspects in product standards*. This document will be a guide for standard writers in areas outside environmental management, and ISO is giving the document a final review before publishing it as ISO Guide 64.

In Oslo, TC 207 continued to move ahead in all its work areas; in addition to the new DISs, progress was made in the documents being developed for environmental labelling, environmental performance evaluation, life cycle assessment, and terms and definitions.

In environmental labelling, standards are being developed for three types of labelling programmes. Type I labelling programmes are called "practitioner" programmes, that is product- or product category-based labelling programmes such as Canada's Environmental Choice Programme or Germany's Blue Angel Programme.

The type I standards will provide a common approach for these programmes. Type II labelling programmes are based on common terms and definitions which are then used for self-declared claims. Type III programmes are based on a "report card" concept, much like existing nutrition labels. The labelling series will also eventually include a standard outlining a set of principles common to all types of labelling programmes.

In the area of environmental performance evaluation, a working draft of a guideline has been developed. This document will provide an organization with guidance on how to develop and implement an ongoing internal evaluation system. It will also provide guidance on the development and selection of performance indicators.

In life cycle assessment, several standards are being developed. They deal with each stage of the environmental life cycle assessment of a product, including inventory assessment, impact assessment and improvement assessment.

The actual publication of five documents was a first for TC 207 and marks a new stage in its development. With these documents published, TC 207 and several of its sub-committees are taking steps to establish and strengthen links with organizations that will use or reference standards in the ISO 14000 series. These include the European Committee for Standardization (CEN), the World Trade Organisation (WTO), ISO/TC 176 in quality management systems and CASCO, the ISO Council Committee on Conformity Assessment. The latter organization is likely to be involved in maintaining international consistency for EMS registration schemes.

PRACTICAL, USEFUL AND USABLE

Standards as an instrument of business strategy must provide the sort of flexible guidance that enables innovation through the systematic development and achievement of continually improved policies, objectives and targets. This may sound like a given but it is not an easy goal to achieve.

A large percentage of the world's economy is in the hands of small and medium-sized enterprises (SMEs). In Canada, for example, 82% of companies have less than 200 employees and they produce 43% of Canadian GNP. I am sure that similar statistics exist for Korea and for other countries in that region. The needs of lesser developed countries (LDCs) and newly industrialized countries (NICs) must also be considered. A flexible system has to be responsive to resource issues, expertise and capability issues and cultural issues. The standards must be practical, usable and useful. They must be easy to understand. They must produce benefits and not just be a drain on resources. And yet, they must provide scope for the state-of-the-art systems and improvements that large transnationals demand and are capable of.

I would suggest that to do this there must be a gradualist approach. An organization must be able to enter the system at a level of performance of which it is capable. This must then be supported by the principle of continual improvement. To establish a set of minimum requirements that SMEs and LDCs cannot possibly meet is to serve only the needs of a limited constituency and would reduce the practicality and usefulness of the documents. It would be to create a new set of barriers to market access.

There must be practical guidance offered and not just a series of criteria that constitute hoops to jump through. The objective is to improve business and environmental performance. This means mobilizing these standards effectively; it means getting people to use them. The best test of their success will be their acceptance by the marketplace. If they are full of brilliant state-of-the-art thinking but sit on the shelf because nobody can figure them out or use them, they will have failed.

THE CONFORMITY ASSESSMENT QUESTION

These points lead to the important question of conformity assessment. The issues of certification and registration of environmental management systems are not easy to resolve. Will registration programmes keep SMEs and developing countries out of certain markets? Will they require unwanted disclosure of performance information? Will they be *de facto* mandatory? Will their cost be prohibitive? Will there be a proliferation of different management systems registrations? Will they become a trade barrier?

If EMS certification or registration is going to become a fact of life, then workable, cost-effective alternatives must be developed that provide opportunities to participate. Serious consideration should be given to the idea of self-declaration. The credibility and acceptability of internal audits and reviews must be defined and understood. And mechanisms for mutual acceptance and recognition must be made a high priority. The goal is not to create exclusionary barriers but to enable, and to provide assurance and recognition of performance and improvement. A system that is or appears to be "exclusionary in spirit" will cause organizations to react negatively to these standards and to opt out of participation.

While the standards in the ISO 14000 series are all voluntary standards, they are designed to enable the provision of assurance of performance through audits. Many organizations may decide to implement an environmental management system with the help of all or parts of ISO 14001 and ISO 14004 and then rely on internal, first-party audit programmes. Many others, however, because of pressure from customers or the public, may decide to have their environmental management system registered through a third-party audit.

For registration purposes third-party audits must be conducted by accredited registration bodies. It is expected that a requirement for accreditation programmes will be the use of certified environmental auditors. To use a Canadian example, the Standards Council of Canada accredits registration bodies for the ISO 9000 series of standards and is the body that will also accredit registration bodies for the ISO 14001 standard. Thus accredited registration bodies such as the Quality Management Institute (QMI), a division of the CSA, will register an organization's EMS in the same way that an organization can now be registered to ISO 9000. Exactly who will certify environmental auditors in Canada is still not decided. However, discussions are currently underway with various interested parties and a resolution is expected soon.

Before the release of ISO 14001 in 1996, many companies around the world elected to register to BS 7750 as an interim approach. Other organizations took steps to register to the committee draft (CD) ISO CD 14001 and still others used the ISO DIS 14001 draft international standard (DIS).

While registration programmes are becoming established, several countries are taking the initiative to develop EMS pilot programmes to help bring their large international traders up to speed with ISO 14001—with the expectation that registration will become a business imperative after 1996.

One special certification issue with which TC 207 is still grappling is EMAS, the European Union's Eco-management and Audit Scheme regulation. Is it having too much influence on TC 207 and does it have the potential to become a trade barrier? Added to this is the fact that in June 1993 the European Union (EU) gave CEN, ISO's European regional equivalent, the mandate to deliver an EMS standard within 18 months from March 1994. If the ISO process doesn't satisfy the EU, will there be competing standards? Fortunately, the compromises reached in Oslo between the European and non-European countries appears to be salable to EU officials and it is likely that ISO 14001 will be accepted as an alternative to EMAS, with minimal additional requirements contained in a short bridging document.

ENVIRONMENTAL STANDARDS AND TRADE

Exclusionary barriers are an international trade issue. It is worth recalling that one of the stated purposes of the ISO is to facilitate trade.

The basic premise of freer trade agreements, such as the General Agreement on Tariffs and Trade (GATT) and the World Trade Organisation (WTO), is that open market access creates greater efficiencies of production and distribution and a more even and equitable distribution of economic opportunity and benefit. Standards are one of the major fulcrums of these agreements. They can be used either to create local and idiosyncratic rules and hence barriers to access—too often the traditional role—or they can be used to enable compatibility and interchangeability and hence remove barriers. Products and services that are compatible and interchangeable move more freely through world markets than those that are not.

The criticism levelled against finding consensus-based compatibility is that it promotes downwards harmonization. This is especially true in the environmental area where environmental groups, who in general are for more command-and-control regulation rather than voluntary standards, warn of the possibility of reduced performance levels. Business people in developing countries, on the other hand, do not want foreign standards which they cannot meet imposed on them in the name of harmonization. They see this as an infringement of national sovereignty and the creation of barriers to trade. Trade agreements recognize national sovereignty and national treatment. Each country has the right to set its own levels and each country must give the same treatment to the products and services of foreign companies as it gives to its own.

How, then, do you "level the playing field" and make the game fair for all the players without lowering standards on the one hand or creating non-tariff barriers on the other? The answer of TC 207 is to develop a private-sector, assurance-based, systems approach based on the principle of continual improvement:

- *private sector* to avoid the issue of national sovereignty—market forces are **international**;
- *assurance-based*, using appropriate mechanisms and a gradualist approach, because what gets measured gets done;
- *systems-oriented* because we must deal with these issues holistically and strategically, we must encourage and enable innovation and rethinking if we are to achieve the magnitude of change required;
- and based on the principle of *continual improvement* because the approach is gradualist and designed to encourage entry at any level and because the status quo is simply no longer acceptable.

MEETING THE CHALLENGE

The work of TC 207 is well underway with five key documents published in 1996, but it has a long way to go and many challenges ahead. Overriding everything else is the need to develop top-notch standards within just a few years. This creates concerns about respect of process and product quality. Is there adequate participation by developing/newly industrialized countries and by representatives of small and medium-sized enterprises? Is there balanced representation on the national delegations? Are environmental and consumer groups being heard as well as industry and the consultants?

In the final analysis, the success of this endeavour will only be ensured if industry uses the standards. They must be useful and usable. They must enable not only improved environmental performance but also improved business performance, and not just for large transnational companies. They must be accessible to small and medium-sized enterprises and to the businesses in developing countries. They must be usable at all levels of performance and capability.

The ambitious plans of TC 207 called for the first documents to be published in 1996. It is very likely that the ISO 14000 series of environmental management standards will spread much more quickly than the ISO 9000 series of quality standards did. The management systems standard culture is now established, the need is perceived by business and the infrastructure is there to support it, with over 100 000 organizations in 93 countries now registered to ISO 9000 alone.

These new standards will continue to increase the strategic importance of international standards. They will become an integral part of the mechanism of global trade. And by enabling improved environmental performance they will be contributing to one of the key purposes of standards, one that we should not lose sight of: they will be contributing to the public good.

Ultimately industry will only improve its environmental performance when it has a strategic and competitive interest in improving its environmental performance. The more international standards can do to encourage, facilitate and become a focal point for these interests, the more they will be serving the larger interests of the public good.

CONCLUSION

In summary, the environmental trends that have gathered momentum for the past 30 years cannot be stopped, and they should not be ignored. They are growing. Eventually, their influence will touch every corner of the global business community.

Today, the ISO 9000 standards are becoming the standards by which business does business. In a few years, the ISO 14000 standards will play a complementary and supporting role.

The ISO 14000 standards will be voluntary. They will provide business with the tools to meet growing environmental pressures. They are responsive to the potential environmental negatives of modern business.

At the same time, the ISO 14000 series approaches environmental management from a positive perspective. Like the ISO 9000 series, it assumes that the more efficient and quality driven an organization is, the more competitive it becomes.

The ISO 14000 standards will be appropriate for businesses of all kinds, all sizes, all types. It is the key to meeting the environmental demands of consumers, customers, suppliers, bankers, investors and legislators. At the same time, it is an opportunity to improve efficiency and competitive ability.

To prepare for ISO 14001, it is important to have commitment from top management. And it helps to be familiar with process management standards such as those in the ISO 9000 series.

Environmental management standards are systematic. They help a company establish where it is in the present, where it plans to go in the future and how it must meet its own individual targets to get there. Along the way, they help develop communication and reporting systems and performance monitoring systems, so performance is ultimately improved. Environmental management standards will affect all businesses profoundly at some point in the not-too-distant future.

NOTE

It is acknowledged that the basis of this chapter has already been used in various materials produced by the Canadian Standards Association.

REFERENCE

Wilkinson, S. (1991) *Canadian Intellectual Property Review*, Vol. 8

3
An insight into the development and implementation of the international environmental management system ISO 14001

Oswald A. Dodds[1]

INTRODUCTION

The management of an organization—whether private or public, large or small, wherever it is located in the world, whatever the culture and state of development of its host countries, whatever its products or services—needs to be efficient and effective if it is to survive and prosper.

Organizations need clear parameters or rules to guide their personnel and customers. Most people would see such concepts as "systems"—arrangements of parts which work together as a whole, in this case designed to move the organization towards its goals (be they profit, quality, service, customer satisfaction or whatever else is relevant).

There will clearly be many "systems"—for production, sales, marketing, finance, safety and so on. Some will be clear, comprehensive, written and well known by all

Environmental Management Systems and Cleaner Production, edited by R. Hillary.
© 1997 John Wiley & Sons Ltd.

involved in their use. Others may not; they may be handed down from operator to operator, salesperson to salesperson, part of the culture or informality which surrounds some organizations and people. Is this latter approach efficient and effective? Do the organization and its customers benefit? What about new people—how do they find out what goes on in the organization and how they are expected to perform their roles?

I start from a position which opposes such informal approaches to management. I also oppose rigidity and excessive formality. My preference is for written systems, regularly reviewed and updated, written in as plain a style as possible, which are relevant to, known by and used throughout the organization. Practicality and cost are important considerations but, in simple terms, written systems should cover all significant aspects of the organization and its mission (or purpose). It is essential that all elements of such systems can be interlocked (where necessary and appropriate) to achieve maximum effectiveness at minimum cost.

I am also concerned to avoid duplication of effort and to use best practice wherever possible. How can this be achieved? One obvious option is by using and refining other systems. Another is by the use (and adaptation) of a "model", developed perhaps by a trade association or professional, learned or other body. Such "models" exist for many aspects of management and for many other issues. Clearly, the needs and wishes of customers and suppliers, export requirements, legal issues and the activities of competitors must also be considered in developing the business management system(s).

I am led to conclude that there is a need for a systems-based approach to managing any organization and that this is best achieved through the creation and use of a written management system, using a recognized "model" which is accepted at least nationally, preferably on a regional (say continent) basis and ideally at an international level. I also suggest that a series of linked parts, following an agreed common core, will best serve the purpose and be effective and efficient most of the time. In many countries, legal requirements reinforce the need for such an approach.

Given the above comments, I believe that the national and international standards writing bodies (e.g. British Standards Institution (BSI), European Standards Body (CEN), International Organization for Standardization (ISO)) have a major role to play in producing the necessary standards through their consensus-based procedures. "Quality" aspects (the ISO 9000 series of standards) are dealt with by ISO TC 176, and ISO TC 207 was created in 1993 to deal with environmental management tools and systems. Discussions have also commenced and faltered within ISO on the need for integrated business management system standards and when, how and by whom they should be produced.

Standards bodies operate through committees, sub-committees and working groups, each composed of nationally nominated delegates and experts which meet and agree any standards for formal approval by the standards body. The process is traditionally slow and cumbersome—a result of the need to achieve agreement between a wide range and large number of individuals and countries.

ENVIRONMENTAL MANAGEMENT—WHAT IS IT AND WHY IS IT IMPORTANT?

With the increasing worldwide concern about the environment (sustainable development, ozone depletion, destruction of the tropical rainforests etc.) and the actual or potential effects of business, industry and the private citizen on the future of planet Earth as we know it, there is pressure on organizations of all types to manage their impacts on the environment. Managing (for) the environment is a relatively new concept to many in business, industry, commerce and local and central governments throughout the world. Times and attitudes are changing and the environment is also emerging as good (and profitable) business.

Many organizations, perhaps prompted by concerned external bodies (e.g. financial institutions and pressure groups), wish to improve and demonstrate their environmental achievements and performance. Many also do so against the background of increasing awareness of environmental issues and pressure for improvement from customers, shareholders, employees and the public, as well as the increasingly demanding requirements of legislation. These pressures were reflected in documents such as the ICC Charter for the Environment, the Keidandran Principles, Global Environmental Management Initiative (GEMI), the Baldridge Award and many more and, perhaps as a consequence, by requests to standards writing bodies to produce standards for environmental management systems. Many such bodies and (in Europe) regional bodies established advisory groups and/or technical committees to begin the standards writing process. This is how BSI came to create its Technical Committee and how the British Standard BS 7750 came into being. ISO received the same reaction from its members and others and, eventually, TC 207 was created with TC 207 SC1, established to produce internationally agreed environmental management system documents.

TC 207 first met in June 1993 and has a Canadian secretary (from the Canadian Standards Association) and chairman. It recently had 67 ISO members involved in its discussions and 27 recognized liaison bodies. It is working through six sub-committees and two direct reporting working groups. These are:

Sub-Committee 1	Environmental Management Systems
Secretariat	UK (BSI—British Standards Institution)
Sub-Committee 2	Environmental Auditing
Secretariat	Netherlands (NNI—Nederlands Normalisatie Instituut)
Sub Committee 3	Environmental Labelling
Secretariat	Australia (SA—Standards Australia)
Sub Committee 4	Environmental Performance Evaluation
Secretariat	USA (ANSI—American National Standards Institute)
Sub Committee 5	Life Cycle Analysis

Secretariat   Germany (DIN—Deutsches Institut für Normung eV)

Sub Committee 6   Terms and Definitions
Secretariat   Norway (NTS—Norwegian Technical Standards Institute)

Working Group one has a German (DIN) convenor and deals with "Environmental aspects in product standards"—a guide intended for other standards writing groups/committees to remind them to write environmental issues into product and other standards.

Working Group two has a New Zealand convenor and is looking at the use of ISO 14001 and ISO 14004 in the forest industry.

ISO TC 207, SUB-COMMITTEE 1—ENVIRONMENTAL MANAGEMENT SYSTEMS

The chairman and secretariat of SC1 are from the UK (BSI). The Sub-Committee was given three work items to deal with. These are:

- Environmental management systems—specification with guidance for its use.
- Environmental management systems—general guidelines on principles and their application.
- Environmental management systems—guidelines on special considerations affecting small and medium-sized organizations.

At its first meeting (in October 1993) in Amsterdam, the Sub-Committee formed two working groups to produce the "Specification" (WG1) and "Guidelines" (WG2) documents referred to above. The committee also decided to deal with its third work item initially by appointing specialists (from Ireland and Denmark) to focus on small- and medium-sized enterprise (SME) issues and to ensure that SMEs were not overlooked by the other groups. It has now created a task force to examine the issue in more detail.

SC1 also agreed a very ambitious timetable for completing its work.

The Working Groups produced committee drafts of proposed standards which were circulated for comments in September 1994. The comments were considered at meetings in February 1995 and the revised committee drafts were circulated for ballot. The results of the ballots were 93% support for the documents becoming Draft International Standards (DIS). The ISO DIS 14001 and 14004 were formally balloted on, and as full International Standards were formally approved and published in September 1996.

THE SPECIFICATION—CONTENT AND APPROACH OF ISO 14001

The Working Group drafting ISO 14001 worked long hours seeking to agree a consensus-based document. This involved many hours of hard bargaining and negotiations—not only on the need for, or basis of, an ISO environmental management system, but (mainly) on the detailed content of the proposed document. Reconciliation of different regional, legal frameworks and cultural and economic issues proved difficult but significant progress was achieved on all fronts.

The Working Group's intention—the production of a good system—has been realized, without compromise, as the group believes (from my perspective at least) that the approach will achieve benefit and acceptance on an international basis. Several statements have been inserted to emphasize that the intent of ISO 14001 has been to recognize and reconcile regional and national differences and not for the document to produce a rigid (uniform) output-focused system, to create barriers to trade or to concentrate on a system leading to enhanced performance rather than a performance-based system. These were difficult concepts to understand for many experts involved in the SC1 and its working groups and will, no doubt, be difficult to "sell" to others in the future, but the determination to do so exists throughout those involved.

The structure of ISO 14001 is:

- Introduction
- Scope
- Normative References
- Definitions
- Environmental management system requirements
- Informative Annex
  A Guidance on the use of the specification
  B Links between ISO 14001 and ISO 9001
  C Bibliography
- Tables 3.1 and 3.2 showing correspondence between ISO 14001 and ISO 9001 (Reproduced as Tables 3.1 and 3.2 in this chapter)

The Specification's (ISO 14001) environmental management system clause headings are:

- General requirements
- Environmental policy
- Planning
  - Environmental aspects
  - Legal and other requirements
  - Objectives and targets
  - Environmental management programme(s)
- Implementation and operation
  - Structure and responsibility

Table 3.1 Correspondence between ISO 14001 and ISO 9001

ISO 14001: 1996		ISO 9001: 1994	
General requirements	4.1	4.2.1 1st sentence	General
Environmental policy	4.2	4.1.1	Quality policy
Planning			
Environmental aspects	4.3.1	—[1]	
Legal and other requirements	4.3.2	—[2]	
Objectives and targets	4.3.3	—	
Environmental management programme	4.3.4	4.2.3	Quality planning
Implementation and operation		4.1.2	Organization
Structure and responsibility	4.4.1	4.18	Training
Training, awareness and competence	4.4.2	—	
Communication	4.4.3	4.2.1 without 1st sentence	
Environmental management system documentation	4.4.4	4.5	General
Document control	4.4.5	4.2.2	Document and data control
Operational control	4.4.6	4.3[3]	Quality system procedures
	4.4.6	4.4	Contract review
	4.4.6	4.6	Design control
	4.4.6	4.7	Purchasing
	4.4.6	4.9	Control of customer-supplied product
	4.4.6	4.15	Process control
			Handling, storage, packaging, preservation & delivery
	4.4.6	4.19	Servicing
	—	4.8	Product identification and traceability
Emergency preparedness and response	4.4.7	—	
Checking and corrective action			
Monitoring and measurement	4.5.1 1st & 3rd paragraph	4.10	Inspection and testing
	—	4.12	Inspection and test status
	—	4.20	Statistical techniques
Monitoring and measurement	4.5.1 2nd paragraph	4.11	Control of inspection, measurement and test equipment
Non-conformance and corrective and preventive action	4.5.2 1st part of 1st sentence	4.13	Control of non conforming product
Non-conformance and corrective and preventive action	4.5.2 without 1st part of 1st sentence	4.14	Corrective and preventive action
Records	4.5.3	4.16	Control of quality records
Environmental management system audit	4.5.4	4.17	Internal quality audits
Management review	4.6	4.1.3	Management review

1 Legal requirements addressed in ISO 9001, 4.4.4.   2 Objectives addressed in ISO 9001, 4.1.1.   3 Communication with the quality stakeholders (customers).

Source: International Standard ISO 14001, BSI 1996.

Table 3.2 Correspondence between ISO 9001 and ISO 14001

ISO 9001: 1994		ISO 14001: 1996	
Management responsibility			
Quality policy	4.1.1	4.2	Environmental policy
	–[1]	4.3.1	Environmental aspects
	–[2]	4.3.2	Legal and other requirements
	–	4.3.3	Objectives and targets
	–	4.3.4	Environment management programme(s)
Organization	4.1.2	4.4.1	Structure and responsibility
Management review	4.1.3	4.6	Management review
Quality system			
General	4.2.1 1st sentence	4.1	General requirements
	4.2.1 without 1st sentence	4.4.4	Environmental management system documentation
Quality system procedures	4.2.2	4.4.6	Operational control
Quality planning	4.2.3	–	
Contract review	4.3[3]	4.4.6	Operational control
Design control	4.4	4.4.6	Operational control
Document and data control	4.5	4.4.5	Document control
Purchasing	4.6	4.4.6	Operational control
Control of customer-supplied product	4.7	4.4.6	Operational control
Product identification and traceability	4.8	–	
Process control	4.9	4.4.6	Operational control
Inspection and testing	4.10	4.5.1 1st & 3rd paragraph	Monitoring and measurement
Control of inspection, measuring and test equipment	4.11	4.5.1 2nd paragraph	Monitoring and measurement
Inspection and test status	4.12	–	
Control of non conforming product	4.13	4.5.2 1st part of 1st sentence	Non conformance and corrective and preventive action
Corrective and preventive action	4.14	4.5.2 without 1st part of 1st sentence	Non conformance and corrective and preventive action
	–	4.4.7	Emergency preparedness and response
Handling, storage, packing, preservation and delivery	4.15	4.4.6	Operational control
Control of quality records	4.16	4.4.3	Records
Internal quality audits	4.17	4.4.4	Environmental management system audit
Training	4.18	4.4.2	Training, awareness and competence
Servicing	4.19	4.4.6	Operational control
Statistical techniques	4.20	–	
	–	4.4.3	Communication

1 Legal requirements addressed in ISO 9001, 4.4.4.   2 Objectives addressed in ISO 9001, 4.1.1.   3 Communication with the quality stakeholders (customers).
Source: International Standard ISO 14001, BSI 1996

- Training, awareness and competence
- Communication
- Environmental management system documentation
- Document control
- Operational control
- Emergency preparedness and response
• Checking and corrective action
- Monitoring and measurement
- Non-conformance and corrective and preventive action
- Records
- Environmental management system audit
• Management review

Of major concern throughout the discussions were the possible impact, relevance and consequences of the European Union's Eco-management and Audit Scheme (EMAS) on any ISO standard(s) in the areas of environmental management systems and environmental auditing. EMAS, agreed in June 1993, introduced, on a European-wide basis, a scheme whereby eligible companies which comply with it can be assessed and awarded a "Euro-logo" signifying their participation in and compliance with the scheme. One method of achieving compliance (not the only one) is by certification to EU-approved national, European (CEN) or international standards. Indeed, the European Commission has "mandated" the European Standards Body (CEN) to produce the necessary European standards. Clearly, such a mandate places CEN and its members under an obligation which can be achieved (through joint working arrangements known as the Vienna Agreement) at ISO level, but the level of detail required by EMAS and the CEN mandate's timetable place extra pressures on European members of ISO, hence the debate and concerns by European and non-European members alike! CEN has agreed to adopt the ISO standards without change and submit them for approval under EMAS. It must also be remembered that CEN may be (effectively) forced to write its own standards or adapt ISO standards dealing with the subject, e.g. ISO 14001, if the ISO products (i.e. standards and guidelines) cannot be "approved" under EMAS.

THE GUIDELINES—CONTENT AND APPROACH

Working Group 2 of SC1 set itself a similar production target to that of WG1. The Guidelines document, ISO 14004, is guidance based only and has not been prepared with certification (assessment) in mind. It is intended to introduce the reader to the concepts and principles of environmental management systems, to reference publications and principles from around the world and to offer organizations information to help them embark on environmental management. The document has been written to complement the Specification-based document ISO 14001, which will allow organizations to transfer to the latter document with minimum difficulty should they wish to do so.

THE ISO STANDARDS—THEIR INTENDED USERS?

Throughout the preparation and production process so far, the needs and wishes of all types and sizes of organization—the intended users of the standards—have been paramount. These needs have been considered in the context of developed and developing areas of the world and in a way intended to be useful and acceptable to any organization regardless of their type, activity or location. As stated earlier, local conditions and economic factors have been considered and the documents constructed, throughout their development, with these issues in mind. National legal systems—on a worldwide basis—have also been considered, as have the various approaches to the enforcement of legislation and the use of the courts.

Clearly, the experience of some countries in the development and use of standards for environmental management systems has influenced the development process, as have the experiences of practitioners, many of whom are from Europe and the so-called developed and/or industrialized areas of the world. It is however, in my view, a mistake and also premature to equate such experience and input as leading to ISO standards merely applicable in European-style industrial economies. It is also the case that the standards are being written to be applicable to manufacturing, service, large and small organizations of every type—on the worldwide stage.

The authors of the standards are representatives of business, industry, learned bodies, consumers and many other groups and they quite properly reflect the views of their "sponsors" in the process. Practitioners also feature prominently as Working Group members. Documents are only created where there is an accepted need.

I hope that the Sub-Committee/ISO vision that standards are necessary is shared by a wide audience which will assist in the development process and that business will use the emerging documents—early experience suggests that many are doing so.

NOTE

1. The views and opinions expressed are not necessarily the views of Northampton Borough Council, BSI or ISO and its members, and are those of the author only.

4

International voluntary standards—the potential for trade barriers

KERSTIN PFLIEGNER[1]

INTRODUCTION

The development of voluntary standards for environmental management started in 1992 in Britain with BS 7750, was followed by several other European countries and the European Union's Eco-management and Audit Scheme (EMAS) and has now reached an international level. The ISO 14000 series of standards on environmental management currently developed by the International Organization for Standardization (ISO) has the objectives of improving environmental performance in industry and facilitating trade. By providing an international standard for environmental management systems (ISO 14001), ISO is aiming at achieving a harmonization of such national and regional standards.

The difficulty of setting a worldwide standard is that it has to be applicable in countries with widely different levels of economic development and environmental regulation. Given both economic and regulatory differences, in particular between developed and developing countries, and the fact that the latter are underrepresented in the process of setting the ISO 14000 standards, concerns might be raised that the standards could lead to the requirements and management systems of advanced industrial nations being imposed on developing countries. The latter are not sure whether the industrial sector in developed countries will adopt environmental management systems (EMS) on a wide scale and whether these policies will restrict or facilitate their exports.

A recent survey conducted by the United Nations Development Programme (UNDP) among experts involved in the ISO 14000 standard setting has shown that there is uncertainty whether the standards will create non-tariff trade barriers or not.

Environmental Management Systems and Cleaner Production, edited by R. Hillary.
© 1997 John Wiley & Sons Ltd.

One of the main arguments against ISO 14001 being discriminatory is that certification to the standard is voluntary. As a voluntary scheme, the standard cannot create an official trade barrier as defined by the World Trade Organization (WTO) in its Technical Barriers on Trade Agreement (TBT). However, the effect of ISO 14001 on developing country trade will depend on the status that the standard gains in conducting business. The wide implementation of ISO 9000 series of standards has proved that, although the standards are voluntary, they have become a requirement for supplying organizations worldwide. The same is expected to happen with ISO 14001. At present, governments in industrial countries are considering the role of this standard within their regulatory system. In the European Union, ISO 14001 certification may satisfy the EMS requirement of the EMAS regulation.

As also discussed by United Nation Conference on Trade and Development (UNCTAD) (see UNCTAD 1995) and United Nations Industrial Development Organisation (UNIDO) (see Luken et al in ISO/CASCO 1995) ISO 14001 has the potential of both positive and negative trade impacts. Requirements for ISO 14001 certification, whether demanded by marketplace or government, can disadvantage producers in developing countries which, for various reasons, have difficulties in becoming certified. Those producers may suffer a loss of competitiveness or even market access. On the other hand, positive trade effects are possible for those developing-country firms that achieve certification. The potential of ISO 14001 to create non-tariff trade barriers is analised in the following sections, as is whether firms in developing countries are confronted with special difficulties in obtaining EMS certification and how such problems could be mitigated.

POTENTIAL POSITIVE EFFECTS ON TRADE

Exporters in developing countries could benefit if ISO succeeds in harmonizing national and regional EMS standards. Obtaining information about one international standard is easier than finding out about several unilateral standards valid in different trading countries. One international standard avoids conflicting requirements, reduces costs for multiple inspections, and lessens the complication that companies in developing countries have to be assessed for conformity by certification bodies in each importing country.

The fact that ISO 14001—in contrast to the European Union and British schemes—does not include an absolute performance component might have, apart from possible other shortcomings, positive effects on developing country trade. Certification may be easier to achieve since a company is allowed to set its own environmental performance objectives based on the national regulatory system.

Certification of a company's EMS may increase credibility with buyers, financial institutions, insurance companies, regulators and consumers. Certified companies might achieve increased competitiveness, faster and wider market access, improved market share and higher export earnings. Participation of developing countries in ISO

14001 may increase the rate of foreign investment via improved confidence in local management capabilities.

However, according to a survey conducted by UNIDO among industry associations and standardization bodies in developing countries, non-compliance with ISO 14001 is perceived by the majority of organizations as a threat to the competitiveness of local companies and is likely to impose a barrier to trade (see UNIDO 1995a).

POTENTIAL BARRIERS TO TRADE

There are provisions within the ISO 14001 international standard which have the potential to create trade barriers. A company is encouraged to consider the environmental impact of products when defining its environmental objectives and targets. Products not compatible with those goals might be excluded. Furthermore, procedures and requirements with respect to environmental aspects identified by the company should be communicated to suppliers (see ISO 1996, ISO 14001, subclauses 4.2.a and 4.4.6.c). Large companies in industrialized countries might put pressure on their suppliers, including those in developing countries, to become third-party certified as a means of improving their own environmental performance and demonstrating their environmental responsibility. The pressure could go as far as using certification as a criteria to award preferential trade status, fix supplier quotas or even drop suppliers without certification in favour of certified competitors. This has been the experience with the ISO 9000 series. Certification to this quality standard has often been a requirement for suppliers to maintain trade relations, although registration for suppliers is not a mandatory standard requirement.

Even if companies in developing countries do not have to become ISO 14001 certified themselves, they may have to consider at least certain EMS requirements, if certification spreads among firms in developed countries. The reason is that the life cycle approach underlying EMS considers the environmental performance of suppliers and contractors and this may lead to substitution of inputs or the placing of special requirements on developing country production processes. The fact that an ISO 14001 EMS is company based may force suppliers to comply with different requirements for each company to which they sell products. The difficulties of meeting those requirements increase if they are conflicting or if environmental criteria are only of secondary importance in the supplier country (see UNCTAD 1995).

Producers in developing countries might consider the standards a barrier to trade if they face problems in obtaining certification or fulfilling the environmental requirements of their customers. There are a number of likely obstacles on both the policy and enterprise level, which are described in the following sections.

Lack of participation and information

In contrast to national or regional standard setting, the process of developing international standards by ISO is open to participation by every country with an ISO member body. Although the number of developing nations that are P-members (have the right to participate and to vote on draft documents) of TC 207, the committee working on the ISO 14000 series, increased to 27 by September 1996, most countries report that they are unable to participate effectively in standardization activities.[2] Due to insufficient financial resources, they cannot send representatives regularly to ISO meetings. The lack of active participation in international standard setting deprives developing countries of the possibility of articulating their interests and influencing the outcomes. Standards, rules and procedures are mainly set by industrialized countries and representatives from large companies, whose perspectives might be different. For countries that are not ISO members, draft standards are not accessible. These countries are disadvantaged in so far as they only have access to the standards when they are published and consequently they lag behind nations involved in the process of preparing for EMS certification.

Lack of understanding and expertise

According to the UNIDO survey, lack of full understanding of EMS standards by both managers and government officials is one major obstacle to developing country participation in ISO 14001. Misunderstanding of requirements and wrong usage of ISO 14001 might be caused to some extent by the way in which the standard is written. It leaves room for interpretation by users and needs to be made more concrete before it can be applied in practice.

A lack of expertise, based on a shortage of qualified auditors and consultants, is a further obstacle for developing countries. Having been less exposed to environmental regulations, developing country firms may not have the necessary experience to implement an EMS. Many firms might be unable to fulfill necessary structural and operational changes and to maintain an internal auditing team that meets the standard's requirements without support and training.

Lack of infrastructure and credibility

Third-party certification might be a barrier for developing countries because of a lack of credible domestic certification and accreditation bodies and the costs associated with such international bodies. Although self-certification is an option allowed under ISO 14001, its acceptability to clients remains to be seen. Lack of funding and specialized skills are the main reasons for inadequate domestic infrastructure. Although ISO 14001 certification might not require expensive laboratory or metrology equipment, it will in many cases require consideration of quality management and health and safety standards, which themselves might necessitate a more elaborate infrastructure. The UNIDO expert group identified that many developing countries, least developed countries in particular, lack government policies appropriate for the development of a credible certification infrastructure. Governments in these countries often do not emphasize the development and strengthening of institutions to promote implementation of EMS (see UNIDO 1995b). A lack of or poorly implemented environmental legislation deprives firms in developing countries of a basis on which to formulate an environmental policy and to fix objectives and targets.

Even with existing certification infrastructure in developing countries, exporters may face obstacles if foreign trade partners do not have confidence in certificates issued by local organizations. With ISO 9000 standards, importers in industrialized countries often requested certificates from reputable foreign or international bodies. The credibility of a certification system depends largely on the competence of the auditors who carry out the assessments. In the absence of an international system for qualifying auditors, developing countries have to obtain this expertise by enrolling in training courses from reputable organizations abroad. The problem with gaining credibility is that there is a strong reputation effect involved. The credibility of a certification body might be considered differently by various trading partners and is to some extent subjective.

Lack of management commitment

Lack of government incentives, insufficient information and lack of awareness are possible reasons for the low commitment by the management of developing country firms to implementing EMS. Additionally, the Confederation of Indian Industries (CII) identifies as likely barriers to management commitment: inability to realize the benefits of EMS, the perceived complexity of the ISO 14001 standard, confusion about the compatibility of ISO 9000 standards and ISO 14001 and the impression that implementation of the standards are associated with creating too much paper work (see CII 1995). In Mexico, experience with ISO 9000 has shown that the limited attention traditionally devoted to procedures, documentation and records represents an obstacle to ISO 9000 implementation (see International Environmental Systems Update 1995). In Nigerian companies a preference for individual decision making is often combined with a lack of clearly defined organizational structures and lack of training of employees. Manager tend to prefer the absence of any system at all, because systematic measures are in many aspects constraints on individual decision making or might interfere with malpractice, corruption or decision making according to other priorities such as personal requests from superiors, religion, tribe or area of origin. Lack of motivation by employees and even negative reactions to audits and performance requests are identified as additional obstacles in both India and Nigeria (see CII 1995a and Abalaka 1995).

Lack of technology

Experience with the ISO 9000 series has shown that developing countries face certain difficulties with regard to its implementation due to not having the technologies required by the industry to comply with the ISO standards (Barrera 1995). Although implementation of an ISO 14001 EMS does not directly require certain technical equipment, complying with regulations and continually improving environmental performance might not be achievable without investing in new technologies. Clean technologies may not be available in developing countries and may have to be imported at higher costs. The purchase of new technology does also involve costs for training to upgrade qualifications of employees. The investment in new technologies might be not economically feasible, if exporters have to raise their prices to uncompetitive levels. Colombian industry reports cases where foreign buyers have requested specific environmental standards and then backed down from the purchase because of the higher price of the product (Barrera 1995).

Costs involved

According to the results of the UNIDO survey, the high costs involved in participating in ISO 14001 are one of the main reasons which make developing country participation difficult. High costs for consultants and for obtaining and maintaining certification are posing particular problems for smaller firms. An estimate based on ISO 9000 experience shows that a small company that has no environmental programme and no quality system in place might have to calculate about US$90,000 for consultant fees, $20,000 for registration costs and $10,000 for registration maintenance costs which occur every six months.[3] In addition, costs arise in connection with analysis, documentation and auditing of the EMS and the training of employees. A lack of credible local conformity assessment infrastructure in developing countries might force producers to seek registration by overseas registrars and to contract consultants from abroad to gain the necessary expertise.

Small and medium-sized enterprises

The difficulties of complying with ISO 14001 will be most serious for small and medium-sized enterprises (SMEs) in developing countries. Lack of information, lack of qualified staff and the costs involved in implementing and certifying an EMS are especially critical for SMEs.

Being sensitive to difficulties for SMEs, experts within TC 207 made investigations on SMEs and their capacity to implement EMS. The conclusion was that no separate standard was needed for SMEs and that ISO 14001 is applicable for companies of all sizes. TC 207's decision being in contrast to the need for a special SME guideline expressed by delegates from developing countries, was recognized during the last plenary meeting in June 1996. A project group was formed to investigate if

SMEs encountered difficulties in implementing ISO 14001 and to report to TC 207 to decide whether future steps have to be taken.

STRATEGIES TO AVOID TRADE BARRIERS

In order to mitigate the difficulties for developing countries, a dual approach seems necessary, coordinated so as to achieve the objective of industrial adoption of ISO 14001 and availability of credible certification infrastructure. The contribution of industrialized countries is important in order to avoid potential trade barriers associated with ISO 14001. A number of strategies are proposed below.

Harmonization and non-discriminatory application of standards

In order to accomplish the positive trade effects achievable by harmonization of national standards, it is necessary for countries to use the ISO 14001 international standard as a basis for the development and adjustment of national standards.[4] All WTO member countries must carefully control their activities on standard setting and conformity assessment with respect to the Code of Good Practice defined in the TBT. This code lays down practices for the preparation, adoption and application of standards that prevent them from creating unnecessary obstacles to international trade. Furthermore, the mandate of the WTO has to be clarified before deciding whether and by what means WTO could react if unilateral measures in the area of EMS are not justifiable or are discriminatory. If environmental management is used as a technical barrier to trade, steps must be undertaken by aggrieved parties to contest the actions of offending parties. Supported by its member bodies, ISO could function as an impartial authority and offer experts, data or other inputs to WTO in the case of disputes (Navarrele in ISO (CASCO 1995)).

The planned establishment of Quality System Assessment Recognition (QSAR) by ISO, an evaluation body for national accreditation bodies, could serve the interest of developing countries in working towards a credible certification infrastructure. In addition, international guidelines, providing clarification of the requirements of ISO 14001, an international uniform system of qualification of auditors and a code of ethics for auditors and registrars could serve as a basis for harmonized and more objective accreditation and certification procedures.

Infrastructure development

There is a need in developing countries for financial and technical assistance to build up the local infrastructure necessary for EMS certification. The establishment of local certification infrastructure should follow international guidelines to ensure its credibility. If it has a standards information centre, a country could become a member of

ISONET, a worldwide network of such centres developed by ISO to provide access to information about the standards and certification procedures in different countries.

Regional cooperation between developing countries in building up infrastructure could overcome the problem of limited financial resources. The Latin American Integration Association (ALADI), for example, promotes the establishment of a cooperation system to strengthen standardization and certification institutions in 11 Latin American countries in order to reduce dependence on extra-regional organizations.[5]

Increased information and participation

More active participation in the standard-setting process would allow developing countries to gain the necessary information and advance their interest, and could diminish any psychological barriers that may arise from the perception that the standards are imposed by industrialized countries. Within its assistance programme ISO awards travel grants for delegates from developing countries to one or two meetings in the hope that this will generate broader participation. For the future, this programme should be extended.

ISO should allow non-member countries access to draft standards in order to minimize possible negative trade impacts arising from a time lag in adjusting to standards' requirements. ISO should ensure that international standards are user friendly for developing countries, in particular for SMEs, and that publications are well disseminated. Developing countries, on the other hand, should guarantee that information gained from ISO is effectively used and distributed within the country.

Awareness raising and training

Measures to raise awareness of the importance and potential benefits of ISO 14001 contribute to an increased commitment on the part of both government and industry in developing countries. In some countries the role of government standards bureaux, which typically have functioned more as regulators on industry, has to change. Involvement of the private sector in planning and implementing standards and in providing training and certification is very important.

There is a strong need for training and capacity building related to EMS in developing countries, especially if they have export-led strategies. Training should focus on representatives from governments, local training and certification bodies, consultants and business leaders. Assistance could also include teaching material and curriculum development. Financial and technical assistance should focus on SMEs in particular, since they will have the largest need for support. Companies that decide to implement ISO 14001 should receive further assistance and implementation facilitation.

Seminars and fellowships for specialized training of individuals are part of ISO's technical assistance programme for developing countries. ISO also provides developing countries with information on training conducted by ISO member bodies in OECD countries. For the future, ISO, in cooperation with other institutions, should initiate efforts to construct a system by which qualified consultants, trainers and auditors in environmental management could be accessed by all countries in need of their expertise.

Technology transfer

Small-scale producers, in particular, will need technical and financial assistance in order to obtain technologies adequate to comply with environmental legislation and to achieve conformity with ISO 14001. The facilitation of technology transfer serves the needs of producers in developing countries. Governments in those countries should proceed with the liberalization of their economies to attract an inflow of clean technologies. This inflow could be supported by the institution of intellectual property rights, assuring their owners that their rights to transferred technologies are protected. Industries relocating to countries with lower production costs will bring new technologies. However, developing countries have to develop control mechanisms to ensure that the technologies transferred are clean ones (Navarrete in ISO/CASCO 1995).

Support within the private sector

The private sector itself plays an important role in avoiding potential trade barriers which may arise from ISO 14001. First, certified companies in industrialized countries demanding compliance with specific environmental targets on the part of developing country suppliers should take the economic, social and environmental conditions of the supplier country into account. Second, they should provide suppliers in developing countries with sufficient time and advance information to adjust to new environmental requirements. Third, purchasers in industrialized countries can help raise the performance of suppliers with whom they have significant trade relations. Large companies in environmentally sensitive sectors in industrialized countries have experience with environmental issues and are a valuable source of expertise for smaller firms in developing countries. They should provide consultation and assistance aimed at the transfer of clean technologies and know-how, e.g. in the form of "visiting firm consultants" or experience-sharing workshops.

Cooperative strategies within the private sector, especially among SMEs, should be promoted and supported. Business associations and networks as well as chambers of commerce can help business identify the relevant environmental regulations and ongoing changes. The Confederation of Indian Industries (CII) provides an example of an active business network. In addition to awareness raising on ISO 14001, the CII is organizing training and conducted a recent workshop on EMS in cooperation with the International Chamber of Commerce (ICC). These activities are accompanied by a planned project on EMS certification in a cluster of small-scale firms. The purpose of this project is to demonstrate EMS implementation and actively to support selected small companies in achieving certification. The demonstration project will serve as a basis for the development of an EMS guide for SMEs, which has not yet been provided by ISO.

CONCLUSION

The ISO 14001 international EMS standard, although voluntary, is expected to become a binding business requirement in certain industry sectors. Increasing implementation of EMS in industrialized countries might place demands on suppliers in developing countries which they could find difficult to meet. Whereas certification to ISO 14001 could have positive trade effects for developing country producers, companies that are not certified may suffer adverse trade effects. Producers in developing countries—small firms in particular—may find the high costs involved in implementing and certifying an EMS prohibitive. A lack of information, technology, expertise and credible certification infrastructure are further obstacles faced by developing countries.

In order to avoid potential trade barriers, developing countries need to participate more actively in the preparation of international standards such as ISO 14000. At present, the voluntary standard setting processes are largely dominated by Northern governments and producers. Developing countries will need technical and financial assistance in order to build up an infrastructure that allows them to participate effectively in the ISO 14001 scheme. Small firms in developing countries in particular could benefit from cooperative arrangements within the private sector. Support to suppliers in developing countries by their trading partners in industrialized countries could also help.

As stressed by the WTO in its TBT agreement, the ISO has an important role in creating international standards and facilitating global trade. Only if WTO member countries adopt ISO 14001, avoid conflicting requirements and carefully control their conformity assessment activities can the objective of international standard setting be achieved.

NOTES

1. The paper reflects the views of the author and not necessarily those of UNDP.
2. Additionally, 11 developing countries participate as observer-members. With a total participation of 38, developing countries represent more than half of all TC 207 members (51 P- and 19 O-members). In 1995 ISO member bodies existed in about 85 countries, among them 46 from developing nations. See: ISO database and ISO (1995).
3. The consultancy costs are calculated for an American firm, based on an estimated time frame of two months @ 22 working days @ US$1800 per day, plus travel costs, hotel etc. In developing countries the consultancy costs are lower. See UNDP 1996.

4. In this context, the decision of the European Union whether it will recognize ISO 14001 under EMAS is important since EMAS is site based and only open to certification for companies with an industrial site in the territory of the European Union. Participation by producers in developing countries without an industrial site in the EU is excluded.
5. These countries are Argentina, Bolivia, Brazil, Chile, Colombia, Ecuador, Mexico, Paraguay, Peru, Uruguay and Venezuela. See ALADI 1995.

REFERENCES

Abalaka, J.A. (1995) Director General, Standards Organization of Nigeria, Statement presented at the UNIDO Expert Group meeting to discuss the potential effects of ISO 9000 and ISO/DIS 14000 on the Industrial Trade of Developing Countries, 23–25 October 1995, Vienna, Austria.

ALADI (1995) *Technical Barriers to Trade within ALADI*, statement presented at the UNIDO Expert Group meeting to discuss the potential effects of ISO 9000 and ISO/DIS 14000 on the Industrial Trade of Developing Countries, 23–25 October 1995, Vienna, Austria.

Barrera, X. (1995) *ISO 9000–ISO 14000 and Colombian Ecolabelling Scheme*, statement presented at the UNIDO Expert Group meeting to discuss the potential effects of ISO 9000 and ISO/DIS 14000 on the Industrial Trade of Developing Countries, 23–25 October 1995, Vienna, Austria.

CII (1995) *Role of EMS—CII*, statement presented at the UNIDO Expert Group meeting to discuss the potential effects of ISO 9000 and ISO/DIS 14000 on the Industrial Trade of Developing Countries, 23–25 October 1995, Vienna, Austria.

CII (1995a) *Status of Implementation of EMS in India*, Study Meeting on Environmental Management Systems and Control—ISO 14000, Environment Management Division, Confederation of Indian Industries.

International Environmental Systems Update (1995), An ISO 14000 Information Service, CEEM Information Services, November, Vol. 2, No. 11.

ISO (1996) *ISO 14001: Environmental Management Systems—Specification with guidance for use*, Geneva, International Organization for Standardization.

ISO (1995) *Membership*, Geneva, International Organisation for Standardisation.

ISO/CASCO (1995) *Proceedings, Conformity Assessment for Environmental Management*, ISO/CASCO–ISO TC 207 workshop, Geneva.

UNCTAD (1995) *Newly Emerging Environmental Policies with a Possible Trade Impact: A preliminary discussion*, Geneva, United Nations Conference on Trade and Development.

UNDP (1996) *ISO 14000 Environmental Management Standards and Implications for Exporters to Developed Markets*, Private Sector Development Programme, New York, UNDP.

UNIDO (1995a) A Survey on Trade Implications of International Standards for Quality and Environmental Management Systems (ISO 9000/ISO 14000 Series), Technical Report, September 1995, Vienna, Austria.

UNIDO (1995b) Expert Group Meeting on the Potential Effects of ISO 9000 and ISO 14000 Series and Environmental-Labelling on the Trade of Developing Countries, Report, 23–25 October 1995, Vienna, Austria.

5
The possibility of cleaner production worldwide

SYBREN DE HOO

INTRODUCTION

The Cleaner Production Programme was launched in 1989 in response to a decision from the United Nations Environment Programme (UNEP) Governing Council on the need to reduce global industrial pollution and waste.

The objectives of the programme are to:

- increase worldwide awareness of the cleaner production concept;
- help governments and industry develop cleaner production programmes;
- foster the adoption of cleaner production;
- facilitate the transfer of cleaner production technologies.

To meet these objectives, the programme focuses on the collection and dissemination of information on cleaner production, and on developing local skills to:

- explain the concept;
- illustrate technical applications;
- help people develop cleaner production programmes.

These efforts, initiated through a number of different activities, have cultivated an ever-expanding informal network of cleaner production experts, both in the public and private sectors.

Programme activities include:

- information exchange through publications and other written information, as well as through a computerised database (International Cleaner Production Information Clearinghouse – ICPIC);

Environmental Management Systems and Cleaner Production, edited by R. Hillary.
© 1997 John Wiley & Sons Ltd.

- networking of experts in various industry sectors through working groups;
- awareness raising and training;
- support to demonstration projects.

The Cleaner Production Programme is developing as a worldwide cooperative effort involving other UN agencies, particularly the United Nations Industrial Development Organization (UNIDO) and UNEP. It also involves industry and industry associations and non-governmental organizations (NGOs). Industry is indeed the key player in making cleaner production happen. However, governments also have an important catalytic role to play, by creating the regulatory and institutional framework needed to incite industry to choose the road towards cleaner production.

The Cleaner Production Programme was strongly endorsed in Chapters 20 and 30 of Agenda 21, and since then has been included in the recommendations adopted at the Commission for Sustainable Development (CSD) in May 1994. It is within this framework that the Organization for Economic Co-operation and Development (OECD) and UNEP jointly organized, in 1993, a workshop on "strategies and policies for cleaner production", attended by experts from various parts of the world.

On the basis of the outcome of that meeting, a document on *Strategies and Policies for Cleaner Production* was published by UNEP in 1995. Its purpose is:

- to explain to leaders of government and industry that cleaner production is likely to lead to economic benefits, that there are great opportunities to be seized, and that cleaner production is the best way of fulfilling the requirements of Agenda 21;
- to emphasize the importance of adopting strategies and policies to implement cleaner production;
- to spell out the most effective overall strategy to start a cleaner production programme;
- to describe the toolbox of instruments available for implementing cleaner production policies.

This chapter summarizes the contents of that document.

WHAT IS CLEANER PRODUCTION AND WHY INVEST IN IT?

Over the past 30 years, the industrialised nations have responded to pollution and environmental degradation in a sequence of "ignore, dilute, control and prevent". The logical culmination of this sequence is cleaner production (CP), an activity which combines maximum effect for the environment with substantial economic savings for industry.

Cleaner production means the persistent use of industrial processes and products—designed from their inception to prevent the pollution of air, water and land—to reduce waste, to minimize risks to the environment and human health and to make efficient use of raw materials, such as energy and water.

The key difference between pollution control and cleaner production is one of timing. Pollution control is an after-the-event, "react-and-treat" approach; cleaner production, on the other hand, is a forward-looking, "anticipate and prevent" approach. And we know that prevention is always better than cure!

For companies, this means undertaking an assessment ("audit") of products and processes to be able to identify options for cleaner production. In practice, these options fall into five categories:

- improve management and housekeeping;
- change input materials and substitute toxic materials;
- reuse materials on site;
- improve product design;
- improve process technologies.

Time and again, demonstration projects have proved that major improvements in the environmental and economic performance of companies can be made even without substantial technology changes. To be able to make proper technology and investment decisions such assessment is crucial, as it leads, first, to better planning and selection of new in-process technologies (cleaner technologies) and, second, to a substantial reduction in the number of end-of-pipe technologies required. The five-step process described above also makes it clear that financing cleaner production is different from financing end-of-pipe treatment. The latter will be an investment in well-defined equipment, while CP investment deals with a package of strongly related technology changes in different parts of a production process.

Investing in cleaner production is cheaper than continuing to rely on increasingly expensive end-of-pipe pollution-control technologies. Even if the initial investment for pollution control and cleaner production is similar, over time pollution control costs continue to mount while cleaner production costs level off. It is essential that the benefits of cleaner production and pollution prevention be properly weighed against each other. Usually, cleaner production options are less costly to implement, operate and maintain over time because of reduced costs for raw materials, energy, pollution control, waste treatment and clean-up, and continued regulatory compliance. In addition, the greater environmental benefits can be translated into market opportunities for "greener" products.

Arguments for investing in cleaner production are numerous and have implications for both industry and government. Cleaner production:

- leads to product and process improvements;
- reduces costs related to end-of-pipe solutions;
- saves on raw materials (including energy) and production costs;
- increases competitiveness through the use of new and improved technologies and through enhanced raw materials efficiency;
- improves the health and safety of employees and improves the company's public image;
- ensures compliance with national and international regulations.

It is important to stress that cleaner production technologies do exist, are not patented and are widely available and that the techniques for identifying the technologies needed are well developed. It should at the same time be emphasized that cleaner production only partly depends on new or alternative technologies. As mentioned, it can also be achieved through improvements in management and operations. Numerous demonstration projects, in industrialized and developing countries, have provided the evidence.

CLEANER PRODUCTION, SUSTAINABLE DEVELOPMENT AND THE RELEVANCE FOR DEVELOPING COUNTRIES

The United Nations Conference on Environment and Development (UNCED), held in Rio de Janeiro in June 1992, established new goals for the world community which involve environmentally benign forms of development. Cleaner production is a logical outcome of the sustainable forms of economic development endorsed in UNCED's Agenda 21. Cleaner production can minimize or eliminate the need to make trade-offs between economic growth and environment, between worker safety and productivity, and between consumer safety and competition in international markets. Optimizing several goals at the same time in this way leads to "win–win" situations in which everyone gains.

Cleaner production is now especially attractive to developing countries and those in economic transition because it provides industries in these countries with an opportunity to "leapfrog" over older, more established industries which are still saddled with costly pollution control technologies. These countries can take full advantage of investments in cleaner production while at the same time expanding industrial production.

Finally, while cleaner production is a means of improving industry and protecting the environment, it is equally an effective device for complying with the complex array of rules and regulations designed to protect the environment, such as the requirements for international adherence to conventions on ozone, the discharge of toxic materials, climate change and biodiversity. Industries and nations which embrace cleaner production will find that, in so doing, they automatically fulfil many of their international obligations to the environment.

WHAT ARE THE ELEMENTS OF AN OVERALL STRATEGY TO ESTABLISH A CLEANER PRODUCTION POLICY?

Three important steps are involved in establishing the necessary preconditions for a cleaner production programme:

- establishing a shared vision and consensus that the best way forward is through cleaner production;
- understanding and assessing the existing system, so that areas and sectors requiring change can be easily identified;
- moving forward from assessment to implementation of the initial vision, step by step, through the use of demonstration projects and similar small-scale techniques.

"Start small but think big" may be the best way of summarizing the background of an adequate strategy to establish a cleaner production policy. This approach avoids the common pitfall of trying to do too much too quickly and emphasizes the need to distinguish from the beginning between the short term and the long term. The long-term goal may be to introduce cleaner production to every industry in the country through the establishment of key advisory and information centres on cleaner production and cleaner technologies, thus effecting a nationwide transformation of industrial practice and philosophy. The short-term goal may be simply to set up one effective demonstration project that will start a snowball effect throughout a specific region, industry sector and, eventually, into others.

In establishing a shared vision and consensus about the contents and priority of cleaner production, there is a need to identify the key people and institutions that will form the backbone of the cleaner production initiative. In many cases, one key institution is used to form the nucleus of a "cleaner production centre". This centre plays the role of initiator, raising awareness, through networking and through publications, of the national need for cleaner production. It is not enough to approach just the key leaders in business and government; many, and often quite unexpected, organizations need to be involved. Presentations, lectures, articles, training materials, training programmes, policy studies, policy advice and a cleaner production network—all of these have an important role in establishing a shared cleaner production vision.

Both governments and industries need to have a clear understanding of how their existing system works in relation to cleaner production and if there are obstacles to its implementation which will have to be removed. Obstacles may be conceptual, organizational, technical or economic.

For industries, guidelines for cleaner production opportunity audits are available and have been improved over recent years. Examples include the UNEP/UNIDO *Audit and Reduction Manual for Industrial Emissions and Wastes* and the US EPA *Waste Minimization Opportunity Manual*. Both are translated into several languages, and can be adapted to regional circumstances and used in many countries.

It must be noted that assessment is not a one-off, start-to-finish procedure. It is an ongoing process. Once the assessment has been made and a cleaner production approach adopted, results must be monitored and evaluated. Evaluation will provide feedback to improve the innovations introduced; it will also suggest new areas for the application of cleaner production options. At this point, the assessment cycle begins again.

In carrying out an assessment, it should be borne in mind that there are relatively cheap ways of implementing cleaner production, such as changes in product design, management improvements, good housekeeping, substitution of toxic materials, process modifications, and on-site reuse of wastes.

WHAT ARE THE MAIN ELEMENTS IN THE TOOLBOX FOR CLEANER PRODUCTION POLICIES?

It is up to industry to introduce cleaner production—the role of government is limited to providing an environment that will accelerate the process and encourage industry to initiate its own cleaner production programmes. The range of tools used by governments to facilitate industry adoption of cleaner production is large and still growing. Different countries will naturally choose those combinations of tools which they regard as most suited to their needs and circumstances.

Schematically, the tools fall into four categories:

- adopting regulations;
- using economic instruments;
- providing support measures;
- obtaining external assistance.

In industrialized countries, the first three of these tools have generally been applied in the order given above. In other words, governments have first established regulations designed to limit emissions to the air, water and land. They have then introduced economic instruments to encourage compliance with regulations and to penalize their infringement. Finally, they have provided support for industry to make it easier to meet regulations. However, this approach leads to regulatory systems that are more complicated than they need be.

Regulations have not been introduced on the same massive scale in developing countries, and it is not yet clear whether they will need to be. The implementation of cleaner production, with its ultimate goals of zero emissions and 100% recycling, does not necessarily depend on the existence of an extensive regulatory system. Developing countries may well find that economic incentives for cleaner production, coupled with suitable support measures and the use of external assistance, will be enough to persuade many industries to adopt cleaner production procedures, with regulations playing a less important role than they have in the industrialized countries.

Adopting regulations

Regulations specify, in varying degrees of detail, environmental goals, how the goals are to be achieved and what technology is to be used to achieve them. The more specific, the more rigid the regulation. Cleaner production could be incorporated in any kind of regulation as a leading principle.

A number of lessons can be learned from the experience of many developed countries:

- Although it is important to establish long-term environmental goals, there is a need to allow sufficient time for these goals to be attained.
- Stricter requirements can be made for new industries than for those already established, since the latter have to make larger investments to reduce emissions.
- There is no point in establishing goals if they cannot be implemented and enforced, and if government is not able to ensure compliance. This implies that regulations must be effectively monitored, a process that can prove costly.
- Emission limits often lead to companies adopting pollution control technologies. It is therefore important to specify clearly the goals to which such limits correspond.
- Goals should be defined so that they can be achieved as a result of the adoption of cleaner production and not by the application of pollution control technology.
- Discretionary regulations, which are flexible as to how goals are attained, are preferable to regulations that specify both what must be done and how it is to be done.
- The use of voluntary agreements and codes of conduct, in place of regulations, should be encouraged in different industry sectors. If compliance with such codes is monitored, unnecessary legislation can be avoided. Supporting industry associations is one effective way of stimulating voluntary agreements and monitoring.

Economic instruments

Economic instruments can be used to integrate costs to the environment into the cost of products, and to make the costs of pollution more expensive than the costs of cleaner production. Economic instruments are used to shape and direct technological investments, influence the purchase and use of materials and energy and influence the management of pollution and waste. They can provide incentives for cleaner production or they can, if unwisely fashioned, subsidize pollution control or environmentally harmful industrial activities. There are two forms of economic instruments: those that provide rewards (tax rebates, subsidics etc.), and those that penalize (reporting requirements, taxes, liability, levies etc.). Before applying any of these instruments to promote cleaner production, artificially low prices for resources such as energy and water should be removed. Governments should continuously evaluate the effects of economic instruments to avoid unexpected and unwanted results.

Provide support

Although industry is the prime mover in implementing cleaner production, government support can play a critical role, providing just that bit extra so that industry is persuaded to take the plunge. Government can provide support in four key areas:

- providing information about cleaner technology (including technical assistance);
- assisting in the development of management tools in industry (such as audits, assessment, life cycle assessment (LCA));
- organizing training on cleaner production;
- developing the necessary educational curricula in engineering schools and universities.

As mentioned previously, the establishment of a national cleaner production centre, in the framework of a national environment plan, is an excellent example of institutional support for the development of cleaner production.

International support

Assistance to developing countries or those with economies in transition can take the form of financial aid, transfer of information and know-how, and transfer of hardware. The extent to which the transfer of financial, intellectual and technical resources fosters self-reliance is a key concern in these activities. Capacity building through technology cooperation must enhance the recipient country's ability to manage technological change; it must absorbed as part of the culture rather than as an isolated capability within a government agency or industry. Obviously, financial resources are needed to effect a shift towards cleaner production. Domestic funds and other means of persuasion should be used as a lever to access the obviously more substantial funds available from international loan and aid organizations.

Financial institutions operating internationally must understand the specific character of cleaner production: in the first instance, it requires investment in "software", such as cleaner production assessments, human resource development, launching of well-designed demonstration projects including technical assistance; only later on does it require investment in the implementation of the necessary "hardware". Unfortunately, the rules of many international lending and aid organizations have not provided much scope for investments in the "software" needed to implement cleaner production.

Another problem is that, at the same time as some parts of aid agencies are supporting the transfer of environmentally sound technologies, other parts might be promoting the transfer of technologies without prior assessment of their local or global environmental impacts in the recipient country. In that respect, an environmental assessment of the projects funded by aid agencies would certainly contribute to the promotion of cleaner production.

CONCLUSIONS

Implementing cleaner production requires both the willingness and the means on the part of government, industry and citizens to bring about a transformation of their national economies. This transformation may involve both small and more radical changes which need careful targeting; it requires the phasing in of new policies and initiatives; and it may entail a modification of the roles played by various stakeholders throughout the transition to a cleaner production economy.

The first step must be the development of a shared vision and consensus, and the establishment of the right climate for financial and technical assistance from both national and foreign institutions to facilitate cleaner production. This must be followed up by appropriate strategies and policy instruments. The choice and implementation of measures must be individually fashioned to meet the specific requirements and conditions of a given country. No one formula will apply. The OECD countries are themselves still experimenting with appropriate strategies and policies, and certainly a lot will have to be learned from them.

UNEP will continue to provide a forum for the exchange of experience between all regions of the world.

Section II
National Perspectives

6
Introduction

RUTH HILLARY

This section of *Environmental Management Systems and Cleaner Production* takes many of the issues discussed in Section I and investigates them in the country context to provide an insight into government policies, strategies and initiatives on environmental management systems and cleaner production and the implications these have for businesses. The national perspectives of the USA, Sweden, China, Hong Kong, New Zealand and Australia are presented in the section's six chapters. The common theme in this section is the interrelation between government strategies and policies and the successful adoption of environmental management systems and cleaner production techniques by business, and how these policies can either promote or stifle proactive environmental performance improvements in industry.

In Chapter 7, John Atcheson presents the scenario of US businesses adopting "lean and clean" management and the implications this has for public policy. He draws together the key elements of lean and clean management, pointing out that it is systems thinking married with designing for productivity, efficient use of resources and the involvement of labour that brings about lean and clean companies. These companies, he suggests, outperform compliance-based companies. Furthermore, Atcheson argues that much of US policy is fundamentally flawed because, far from stimulating companies, it discourages innovation, freezes technology *in situ* and encourages investment in single-media, end-of-pipe solutions. He suggests that much of environmental policy and law is influenced by neoclassical economics and therefore based on the premise that environmental performance improvements can only be achieved at a cost. Atcheson argues that US environmental policy needs to change focus from controlling adverse behaviour to encouraging proactive behaviour, citing several programmes established by the Clinton Administration to encourage proactive behaviour. He sets out 10 principles for designing environmental regulations to promote lean and clean practices, finally concluding that an effective public policy needs to recognize the opportunities presented by lean and clean management systems.

Environmental Management Systems and Cleaner Production, edited by R. Hillary.
© 1997 John Wiley & Sons Ltd.

Mikael Backman, in Chapter 8, continues Atcheson's theme of changing public policy but in the Swedish context. He suggests that sustainable development needs to be based on a preventive public strategy involving all citizens along and tangential to the life cycle of a product to develop environmentally adapted solutions. Backman focuses on the shift in Swedish policies which gives preventive strategies a more prominent position citing the grouping of a package of laws into the Swedish Environmental Code. He argues that the code illustrates the trend within Swedish environmental politics and that the changing character of environmental policy has stimulated the emergence of industrial environmental policy largely dominated by larger companies. Industries' response, he suggests, has been to use, although sparingly, a range of environmental management systems tools as well as life cycle assessment (LCA). Utilizing the results of an industrial survey of over 500 Swedish manufacturing companies, Backman discusses the penetration of formalized management systems such as the Eco-management and Audit Scheme (EMAS) and the international standard ISO 14001. He concludes that small and medium-sized enterprises' (SMEs) environmental competence is low but could be stimulated by a range of government mechanisms including environmental taxes or fees on waste.

In Chapter 9, Ya-Hui Zhuang also considers the use of economic incentives as a means to promote cleaner production and environmental management, but in the context of the largest transition economy in the world: China. He describes a national system oriented towards pollution control but with some potential for the successful introduction of cleaner production. However, he suggests that this system is set against a backdrop of outdated facilities, backward technologies and an irrational price structure for water and coal, which all encourage enterprises to pollute and suffer the fines rather than adopt cleaner production solutions.

Zhuang provides an insight into China's cleaner production management, citing the development of ecological planing as part of township enterprises reform, i.e. a multistage utilization of natural resources and wastes which views the township as an ecosystem. He concludes that China's transition to market mechanisms provides the opportunity to promote cleaner production and in the longer term will enable enterprises to cooperate in "build, operate and transfer", capital-intensive cleaner production projects.

Edwin Kon-hung Lui, in Chapter 10, describes the application of cleaner processes in the free-market economy of Hong Kong. He focuses on the introduction of a capital-intensive cleaner production facility designed for hazardous chemical waste from Hong Kong's numerous SMEs. Lui considers the Hong Kong government's attempts to legislate effectively and control chemical wastes which previously ended up in the public sewerage system and the sea. He argues that only by charging for waste disposal at an integrated chemical waste treatment will a clear message go to industry that it needs to consider waste minimization and the "polluter pays" principle. Lui then discusses a range of measures for promoting cleaner processes. He concludes that industry in Hong Kong is becoming aware of the concept of pollution prevention and is gradually replacing end-of-pipe technology with cleaner production.

In Chapter 11, Marje Russ develops Lui's ideas on cleaner production, showing the links between cleaner production and environmental management systems in New Zealand. Russ considers the contextual business and regulatory framework in New Zealand which established the conditions suitable for environmental management systems introduction in enterprises. She argues that New Zealand's unprotected, export-focused market means that its businesses, many of which are SMEs, have become very customer focused considering all their requirements including the environmental ones. This customer focus, coupled with the introduction of an effect-based approach to environmental management in a major piece of environmental legislation, has meant that companies are ripe for the introduction of EMS standards such as the British standard BS 7750 and the international standard ISO 14001. Furthermore, Russ asserts that there are clear and strong links between ISO 14001 and cleaner production methodologies which have been developed for demonstration projects in New Zealand. She concludes that ISO 14001 can provide significant support in promoting the objective of cleaner production.

Brian O'Neill complements Russ's paper by providing an overview, in the final chapter, of cleaner production and environmental management systems initiatives in Australia, suggesting that Australia's leaders and agencies have recognized such initiatives as mechanisms for building an enterprise culture of environmental best practice and innovation. O'Neill argues that international environmental management systems standards can be a model for use in incentive-based legislative approaches to industry designed to reward companies for implementing systems to monitor, control and improve their environmental performance. He adds a caveat, asserting that systems are only as good as their associated environmental objectives and that maximizing an environmental management system's potential requires the inclusion of a continual improvement provision. O'Neill concludes by emphasizing the need to ensure that environmental management systems go beyond mere book-keeping to genuinely promoting environmental protection.

Throughout this section the authors examine a selection of key themes which emerge from the six very differing national perspectives. The shared themes discussed include the following:

- The links between environmental management systems and cleaner production and the potential of environmental management systems to promote and make sustainable cleaner production initiatives.
- National legislative frameworks and policies and their many flaws which promote end-of-pipe technology, and in some cases even encourage companies to pollute, over preventive strategies such as environmental management systems and cleaner production.
- The changing nature of regulatory frameworks which are, to varying degrees, moving towards preventive strategies and the polluter pays principle to stimulate industrial policy and reward enterprises which implement environmental management systems and cleaner production to control their environmental impacts.

- The use and effectiveness of economic mechanisms such as taxes to help promote improvements in industries' environmental performance.
- The way in which large enterprises and customer-oriented businesses are developing effective environmental management systems and in some cases racing ahead of government policy, but also the fact that most enterprises, in particular SMEs, have low environmental competence.

A reccurring theme permeating all six national perspectives is the evolving nature of government policies and national legislation, which seeks effectively and efficiently to improve environmental performance in industry, but sometimes actively works against this objective and individual business efforts. Section III expands and explores this theme by examining the regulatory and self-regulatory mechanisms used to promote environmental performance improvements in businesses.

7
Management systems: getting lean, getting green in the USA

JOHN ATCHESON

INTRODUCTION

In 1982, in the boardroom of Compaq Computer, a quiet revolution was brewing. The new company was getting ready to challenge IBM in the personal computer market. By 1992, when the smoke cleared, IBM—a corporate giant of legendary success—was undergoing one of the most sweeping restructurings in corporate history, and Compaq had established itself as the fastest growing and most successful start-up company on record. The financial details of this revolution are reasonably well known: Compaq achieved the highest first-year sales in the history of US business; by 1985 it was the second largest manufacturer of personal computers; by 1987 it had sales of over US$1 billion dollars, faster growth than any other company in history (Corcoran 1995). By September of 1994, Compaq had moved into first place in home computer sales and IBM had slipped to a distant third, behind another upstart in the computer world, Apple. What is not so well known is that Compaq's and Apple's strategies made them two of the cleanest companies on record.

Compaq's success was tied to what Joseph Romm (1994) calls "lean and clean" management.[1] (See also Porter and van der Linde 1995, Dorfman 1992, Dorfman et al. 1992.) The experiences of Compaq and other lean and clean companies may well settle the debate that is shaping up over just how far companies can go in getting clean and maintaining profitability. And in the process, their experiences can tell us a great deal about how to structure environmental policy and law. The terms of this debate were recently laid out in two articles in the Harvard Business Review. In the

---

Environmental Management Systems and Cleaner Production, edited by R. Hillary.
© 1997 John Wiley & Sons Ltd.

first of these, "It's not easy being green", Noah Wally and Brady Whitehead (1994) argue that so-called win–win strategies for corporate environmental performance—those which make companies greener while improving the bottom line—are rare, and have been vastly overestimated. In the second of these articles, "The challenge of going green" (Harvard Business Review 1994) 12 experts explore alternative views on the subject.

What is perhaps most remarkable about this debate is that, with some exceptions, compliance with regulations is the sole criterion for being green. Romm points out, and the experience of Compaq and numerous others confirms, that when it comes to being green, what happens in the factories and boardrooms is far more important than what happens in legislatures and on the pages of some government's Federal Register.

Quite simply, companies which adopt a strategy of "lean and clean" management, will create less waste and become more productive by making more efficient use of resources, energy, time, capital and labour. Just as simply, government environmental programmes that fail to recognize and capitalize on the opportunities presented by lean and clean management systems will either be suboptimal or, worse, counter-productive.

This chapter will explore the origins and key principles of "lean and clean" management systems, give a few simple case studies of what they have accomplished, examine the implications for public policy and discuss a few pilot programmes that have attempted to encourage adoption of "lean and clean" management in the private sector.

WHAT IS LEAN AND CLEAN MANAGEMENT?

As Romm (1994) notes, Henry Ford pioneered many of the principles of "lean and clean" management nearly six decades ago. Ford focused on eliminating wasted time and wasted resources. Romm shows how these two efforts can be combined in a management paradigm that makes companies more competitive and cleaner.

In 1926, Ford wrote, "We have spent many millions of dollars just to save a few hours' time here and there" (Romm 1994). The cycle time for the Model T Ford from mining to finished car on its way to market was just 81 hours. Five decades later, Taiichi Ohno, one of the founders of lean production and the Toyota production system, wrote:

> To implement the Toyota production system in your own business, there must be a total understanding of waste. Unless all sources are detected and crushed, success will always be a dream.

Most lean production systems have centred on wasted time, focusing on eliminating overproduction, waiting, defective products and excess inventory. What is striking is that the solutions to these production problems read like a pollution-prevention strategy. Just-in-time inventory, facility auditing, total quality management systems and chain-of-custody stewardship, for example, are some of the mainstays of both lean production and pollution prevention. What is even more striking is the power of lean and clean systems that explicitly consider wasted resources as well as wasted time.

Again, Ford, more than half a century ago, had in place a programme which sought to minimize waste. To reduce the use of timber, for example, Ford used the smallest crating for the job; insisted that all crates be opened as carefully as possible so that the wood could be reused; replaced the crates wherever possible with burlap bags or cardboard boxes made from waste paper in Ford's own paper mill; and sent all scrap wood, sawdust, shavings, chips and bark to the salvage department where they were distilled into acetate of lime, methyl alcohol, charcoal, tar, heavy oil, light oil and fuel gas.

But Ford recognized that reuse and recycling were not the best solutions. He practised a lean and clean production paradigm which is being rediscovered today in pollution prevention. As he wrote (Romm 1994):

> It is not possible to repeat too often that waste is not something which comes after the fact. Restoring an ill body to health is an achievement, but preventing illness is a much higher achievement. Picking up and reclaiming the scrap left over after production is a service, but planning so that there will be no scrap is a higher public service.

Ford made prevention almost a moral issue; Michael Porter, in his article "Green *and* competitive: ending the stalemate" (Porter and van der Linde 1995) brings it down to a business level:

> When scrap, harmful substances, or energy forms are discharged into the environment as pollution, it is a sign that resources have been used incompletely, inefficiently, or ineffectively. Moreover, companies then have to perform additional activities that add cost but create no value for customers: for example, handling, storage and disposal of discharges.

ELEMENTS OF LEAN AND CLEAN

A complete description of lean and clean management systems would fill a book. But an analysis of the research done by Romm (1994), Porter and van der Linde (1995), Sue Hall and INFORM (Dorfman et al. 1992) reveals some key elements that have important implications for environmental policy.

Think systems

Incremental decision making and the institutional "stovepipes" that evolve in a bureaucracy are the enemy of "lean and clean" management. Companies which work across boundaries, which encourage integrated and concurrent planning, consistently adopt full-cost accounting, life cycle assessment and tools which make them more proactive, more preventive, more agile.

Romm illustrates this aspect of lean and clean with the following example. Compaq's facilities manager, Ron Perkins, saved the company about a million dollars a year in energy costs. But first, he had to break some corporate barriers. The facilities division had an investment threshold based on a two-year payback period. Perkins knew that the energy-efficient and pollution prevention strategies he advocated would save the company money, but would not clear the investment hurdle. He went to the firm's chief financial office, John Gribi, and made his case. Gribi had a simple philosophy: any investment which met the corporate cost of money plus 3% (at that time about 14%) should get funded. This shifted the return on investment from two years to seven years, saving Compaq millions in operating costs year after year and eliminating emissions into the bargain. "Lean and clean".

Involve labour

A Sealtest ice-cream plant in Framingham, Massachusetts, went from a marginal plant destined for closure to being one of the most efficient plants. Sealtest used on line and worker involvement as the key. A total quality management programme was critical to its success. The plant saved money on energy, reduced emissions and improved productivity by over 10% by reducing the cycle time. According to the utility experts who worked with the company, "The project really took off when the workers got involved."

Design for productivity

People, processes and machines can be made more productive through both prevention and proaction. For example, Romm notes that worker productivity increased by 6% in the Reno post office when lighting was changed. The intent was to improve the post office's environmental performance through energy efficiency. Productivity increases were a bonus. Similarly, INFORM, a non-profit organization, studied 19 facilities which practised pollution prevention. Over 95% of the source-reduction opportunities pursued also increased productivity. The State of Florida has been providing audits designed to identify cost-effective pollution prevention opportunities since the mid-1980s. In the words of the engineer who headed the programme, "I can only think of one plant where we were unable to identify an opportunity to improve the process, save money, and cut emissions." A project conducted by the US Department of Energy identified 75 pollution prevention opportunities with an average return on investment of 63% and a payback time of 1.58 years. The authors of *Dynamic Manufacturing* note that cutting waste by 10% appears to improve total factor productivity by 3%. Again, "lean and clean".

Efficiency of energy and raw materials

The flip side of waste minimization is resource maximization. All too often, companies look only at one aspect of waste minimization, such as reducing toxics or other regulated wastes, and therefore limit the potential for win–win opportunities.

Dow Chemical's Louisiana Division has been running an energy contest since 1982. During that time the price of energy has declined precipitously. Despite this, the average annual return on investment for 575 audited projects comes to 204% and annual savings realized come to $110 million.

As Romm (1994) concludes:

> Companies achieve their dramatic productivity gains by eliminating waste through improving the production process. Focusing on individual operations, what is called industrial engineering in the US, rarely achieves significant or enduring productivity gains. A narrow focus on operations makes it difficult to track down root causes of systemic problems, such as long cycle times, large inventories, high waste, poor communications, flawed strategy, low quality. Yet systemic problems are the obstacles to productivity growth in most companies.
>
> Pollution prevention is the key that unlocks the solutions to many systemic problems ... It forces a company to use many of the same techniques for minimizing wasted resources that the best companies use to minimize wasted time.

Four core principles emerge from the practices of lean and clean companies. First, they emphasize performance-based goals and continuous improvement. Second, they encourage innovation and provide the flexibility and resilience needed to achieve it. Third, they employ accounting practices and data management systems that allow them to see the true costs and benefits of their investments *and* encourage them to value longer-term returns on those investments. And finally, they are anticipatory, seizing and sometimes shaping trends, not simply reacting to them.

THE IMPLICATIONS OF LEAN AND CLEAN FOR PUBLIC POLICY

It seems obvious that if lean and clean companies outperform "compliance" companies in terms of environmental results, it ought to influence public policy. Ideally, environmental policy should be set to stimulate the private policies that result in superior environmental performance. Historically, this has not been the case. In the US, in fact, a restrictive command-and-control system has worked to discourage innovation, freeze technology in place and focus capital investment on the single media end-of-pipe solution; on the incremental rather than the systemic (Atcheson 1996). Understanding why this is so can help us shape environmental policy.

Ultimately, much of our environmental law and policy is based on a simple premise: improved environmental performance can only come at a cost—usually a high one. Much of this muddle-headed thinking, in turn, comes from the pervasive influence of neoclassical economics on policy formulation. To the neoclassical economist, environmental regulation can be viewed as an attempt to internalize environmental externalities. The notion of environment as a *cost-effective business strategy* is simply beyond the view of this paradigm. A quote from Frances Cairncross, environmental economist for the *Economist*, illustrates this misperception:

What the free lunch brigade wants to hear however, is that environmental rules actually persuade companies to take actions that are in their commercial interest but that they had not previously noticed. Remember the economist and his friend who thinks he sees a $10 bill on the sidewalk? "It can't be," says the economist. "If it were there, someone would have picked it up".

Most of the $10 bills to be had by reducing pollution or saving energy have either been picked up already, or can be retrieved only at a cost. *That cost may not be cash but management time. If a bright manager must look for ways to reduce waste output, he or she is not available for developing new markets or streamlining production.* (emphasis added.) (*The Challenge of Going Green*, Harvard Business Review).

This quote embodies many of the fundamental flaws in our public policies and laws designed to protect the environment, and it explains why we presume that protecting the environment will ultimately cost money. A decade ago Amory Lovins faced the same arguments about the potential for energy efficiency. Today, one would be hard put to find a utility without an aggressive demand-side management programme based on his philosophy. The implicit assumption in Cairncross and the expensive lunch brigade's position is that streamlining production and reducing waste are somehow mutually exclusive sets of activities. Thus, the bright manager can only do one or the other.

In fact, "lean and clean" management makes streamlined production systems the method of choice for accomplishing environmental objectives. The mounting empirical evidence belies the neoclassical economist's contention that there is no free lunch. Lovins, Compaq and others are showing that there are a great many "free lunches" out there, and even more inexpensive ones.

The expensive lunch bunch also seems to believe that focusing on environmental performance somehow can only come at the expense of new market and product development. Again, the empirical evidence shows otherwise.

Sue Hall of Strategic Environmental Associates has documented case after case where companies have consciously used environmentally preferable products, processes, and services to capture markets, and in some cases to create them (personal conversation with Sue Hall of Strategic Environmental Associates). Arco's clean fuel and Henkel's phosphate-free detergents are but two examples where companies have done well economically, while doing good environmentally in the market. In fact, as the US's *National Environmental Technology Strategy* and *National Environmental Technology Export Strategy* point out, environmental technologies are the fastest growing global market. And this does not apply just to a few manufacturers of environmental hardware—it applies to our entire economy. Tomorrow's markets belong to companies who can produce goods, provide energy, transportation, shelter, food and fibre, in the cleanest, most resource-efficient manner possible. The demographic trends of population growth, economic growth, and urbanization will create the demand that will make lean and clean companies the winners in current and future markets.

So, there are some free lunches, there are some undiscovered $10 bills. But we need the right set of incentives, policies and regulations to gain them.

BEYOND COMMAND AND CONTROL

There is a general consensus within the US that there is a need to try something new; that the complex, relatively rigid and piecemeal "command-and-control" legal structure is no longer serving the purpose—if in fact it ever did (see Ruckelshaus 1996 and Atcheson 1995). But there is no such consensus on the shape environmental policies should take. One thing is certain: the same demographic trends that are shaping markets must also shape environmental policies. Over the course of the next few decades, humanity will need to make extraordinary leaps in the resource and environmental efficiency with which it delivers products, goods and services.[2] Encouraging and stimulating technological innovation must be the watchword and guiding principle of public environmental policy and lean and clean must be the core of corporate environmental policy.

Environmental policy needs to move from its current focus on controlling adverse behaviour to *encouraging* proactive and "good" behaviour—in short, we need to encourage lean and clean practices.

The Clinton Administration in the US is committed to reinventing environmental policy by increasing flexibility, emphasizing performance not design, while ensuring accountability and better environmental results at lower costs. It has set up several programmes designed to test and pilot new ways of accomplishing environmental objectives:

- *The Common Sense Initiative:* This is aimed at creating cheaper, cleaner, smarter regulations. Pollution prevention, multimedia and facility-wide approaches are being emphasized. The initiative is starting with six key industries: automobile manufacturing, electronics and computing, metal finishing and plating, iron and steel, printing and petroleum refining. Several of the sectors are incorporating EMS into their alternative regulatory strategies.
- *Project XL:* Under Project XL, participants in four categories—facilities, industry sectors, government agencies and communities—are given the flexibility to develop common-sense, cost-effective strategies that will replace or modify specific regulatory requirements, on the condition that they produce greater benefits. Two of the projects accepted, Lucent and Weyerhauser, explicitly focus on environmental management systems and several under consideration propose to do so. Nearly every project selected incorporates some of the principle of EMS.
- *Permits Improvement Team:* The PIT has made recommendations to the Environment Protection Agency designed to streamline the permitting process, identify alternatives to permits, increase the use of pollution prevention and improve the amount and quality of public participation permit decisions.
- *Statutory Integration Project:* This project is examining opportunities to integrate and streamline environmental laws.
- *One Stop Reporting:* Designed to reorient environmental record keeping and reporting to reduce the burden on industry and increase the usefulness and accessability of the data collected.

These programmes together with data from the 22 states with facility planning laws, make it evident that programmes which encourage companies to adopt environmental management systems also encourage them to become lean and clean.

POLICIES ENCOURAGING LEAN AND CLEAN MANAGEMENT

Following are 10 principles for designing environmental regulations for the twenty-first century. They will encourage the adoption of laws and programmes which emphasize environmental management systems as strategies of first choice in the public sector, and will result in lean and clean practices in the private sector.

- *Focus on facilities, communities and ecosystems:* The unit of decision that makes the most sense to industry is not the waste, nor the media, nor the pipe—rather it is the facility. Similarly, the units that make the most sense from an environmental and public health perspective are the community and the ecosystem. Protecting the public from factory emissions of VOCs (volatile organic compounds), for example, while doing nothing about the release of VOCs from transportation and power generation, may not adequately protect people or the environment. We can best understand our problems and measure our progress by looking at the *systems* which are *stressed*—communities and ecosystems. We can best fashion solutions by looking at the systems which create the *stressors*—energy, transportation, manufacturing, agriculture.
- *Emphasize multimedia, multistressor solutions:* We are all familiar with how single-media approaches force suboptimal solutions, but a focus on one release or one waste at a time may be worse. Picture a company faced with a series of air emission limitations for eight different stressors over an eight-year period starting with particulates and ending with CO_2. Initially a scrubber might look like the best bet, but at some point it becomes more costly and less effective than a more systemic solution which addresses all eight stressors simultaneously.
- *Base standards on desired performance, not design:* Although there is a great deal of *theoretical* flexibility in regulations, the practice of specifying "best available technologies" and similar approaches creates a *de facto* design standard which is rarely improved or deviated from in practice by either regulator or the regulated.
- *Emphasize continuous improvement, not bright-line fixed compliance:* While we certainly need enforceable standards, our regulations and policies should be designed to foster a continuous search for opportunities, not just a look at requirements once a decade. Annual facility audits, with goals and indicators of progress rather than a once-a-decade look at a permit requirement, will yield more environmental improvement at lower cost.
- *Expand the use of measurement and feedback:* It would be an overstatement to say—as some have—that for the past two decades we have marked environmental

progress by the number of permits issued, laws written, dollars spent. Nevertheless, the gap between environmental programmes and environmental results has been too large, for too long. Environmental indicators and environmental performance must be the bottom line for environmental programmes. We must improve our monitoring capability, expand the Toxics Release Inventory, and increase the openness of our environmental decision making, to improve our ability to measure results and industry's and the public's ability to engage in dialogue.

- *Increase community involvement in setting goals and evaluating progress:* Once we have meaningful indicators at the facility and community level, we must create more and better ways for local communities to participate.
- *Provide different levels of regulation for different levels of performance:* Companies and facilities which consistently perform better environmentally should not be subject to the same level of stringency as those which perform poorly. Increasing the flexibility and simplifying the administrative requirements we impose on good actors will provide an incentive to improve performance or maintain high standards.
- *Use fiscal tools and market incentives whenever possible:* Concepts such as emissions trading leave the choice of *how* to the regulated community, but preserve or create incentives to improve performance.
- *Regulate at the lowest jurisdiction possible; assign responsibilities to the jurisdiction best able to carry them out:* The current debate on devolution of responsibility for implementing environmental laws is too narrow, both in scope and participation. Arguing about who implements which part of RCRA will not result in much progress. Instead, we need systematic analysis and discussion about the roles of facilities, communities, states, regions, national players and the international community designed to give us a strategic partnership that builds on the strengths of each player.
- *Concurrent and coherent policies:* Traditionally, governments have pursued fiscal, technological, and environmental regulatory policies on separate tracks with little communication or cooperation between agencies. The Administration has developed a comprehensive strategy for technologies necessary to assure a sustainable society.

As noted in *Technology for a Sustainable Future:*

> Our vision is of an integrated system of social, economic, and environmental values that reinforce each other.

If we seek private corporations which work across bureaucratic boundaries, we must have public institutions that do the same. The power of concurrent engineering has been demonstrated time and again. Boeing's 777 aeroplane has come in under cost, ahead of schedule and with fewer redesigns, because it integrated the interests of all its suppliers and customers in the design process. In a similar way, public policies can and must be developed concurrently.

CONCLUSION

An effective public policy needs to recognize the opportunities presented by "lean and clean" management systems in the private sector and must put in place the public tools to encourage companies to become lean and clean. In the US, the trends being pursued were summed up in *Technology for a Sustainable Future:*

> An effective policy framework should look across technologies, across organizations, and across budgets. A narrow focus often misses synergies within our government and between the public and private sectors. The need to bridge institutions is urgent. We need strategies to link scientific, technological, and policy making communities and to join national and international efforts. Ultimately, the development of green technology must become an integral part of all technology policies and programs.

Ghandi, observing his followers heading down the road, said, "There go my people. I must hurry to get in front of them if I am to be their leader." Compaq and other leaders are revolutionizing production through lean and clean management. We in the public sector must heed Ghandi's insight. We need to abandon approaches which freeze capital, fix technology and stifle innovation. As Gustav Speth points out, the time for Luddite strategies is long past. Technology is not only the problem, it is the solution.

NOTES

1. Susan Hall of Strategic Environmental Associates has researched how some companies use environmentally preferable products and processes and green corporate policies to capture market share. Her research will be published this year.
2. Gustav Speth has been making this point for years, and it is the subject of a forthcoming book by Steven Hawkins and Amory Lovins.

    This author conducted a cursory analysis on the environmental effects of the increased use of refrigerators by 2020 *in just China and Latin America*. I used conservative population and economic growth projections and assumed that all refrigerators purchased would be as efficient as the state-of-the-art Golden Carrot refrigerator. I also assumed that power production would be about as efficient as today's average and use a similar mix of fuels. The incremental increase in carbon dioxide emissions from just these two areas, and only from refrigerators, came to more than 8 trillion additional pounds per year.

REFERENCES

Corcoran, E. (1995) "Comeback on a Compaq scale", *Washington Post, Business Section*, 14 September.

Romm, J (1994) *Lean and Clean Management: How to Boost Profits and Productivity by Reducing Pollution*, New York, Kodansha American.

Porter, M. and van der Linde, C. (1995) "Green *and* competitive: ending the stalemate", *Harvard Business Review*, Sept–Oct. pp. 120–134.

Dorfman, M. "Source reduction", *Pollution Prevention Review*, Autumn, pp. 403–414.
Dorfman, M., Muir, W. and Miller, C.G. (1992) *Environmental Dividends: Cutting More Chemical Waste*, New York, INFORM.
Wally, N. and Whitehead, B. (1994) "It's not easy being green", *Harvard Business Review*, May–June pp. 46–52.
—— (1994) "The challenge of going green", *Harvard Business Review*, Jul–Aug. pp. 37–54.
Atcheson, J. (1996) "Can we trust verification?", *Environmental Forum*, Jul.–Aug. pp. 14–21.
Ruckelshaus, W. (1996) "Stopping the pendulum", *Environmental Forum*, Jan–Feb. pp. 25–29.
Atcheson, J. (1995) "If risk assessment and cost benefit are the answer, what was the question?", *Pollution Prevention Review*, Winter pp. 73–83.

8
Swedish national environmental policy and environmental management systems in industry

MIKAEL BACKMAN

BACKGROUND

Efforts by society to improve the environment are in a period of transition between the traditional regulation and control of known problems and efforts based on the idea of preventing new problems before they arise. The "command-and-control approach" to the regulation of pollution has, however, undeniably resulted in lowered emissions from those production facilities subject to this type of regulation. In many cases drastic reductions of certain types of pollution from specific point sources have resulted.

As regulations and other policy instruments have been enacted and put into effect to diminish the environmental impact from the most significant point sources, other problems caused by diffuse sources have become more visible. Diffuse sources, such as the environmental burdens from the use of products and their waste generation, have been increasing in relative importance compared to total environmental impact. The traditional command-and-control regulations, usually designed to solve environmental problems after they have arisen, have limited applicability to this set of problems. An environmental regulatory framework that could make any claims of its ability to remedy all the environmental problems caused through the use of raw

---

Environmental Management Systems and Cleaner Production, edited by R. Hillary.
© 1997 John Wiley & Sons Ltd.

materials, manufacturing processes and production methods, products, patterns of transport and consumption etc. ought to be neither possible nor desirable.

Based on a broad international consensus concerning the goals, the continued efforts to reduce environmental impacts are spreading. Environmental protection authorities, political parties, environmental organizations, researchers and others are, essentially, in agreement that sustainable development must be based on a preventive environmental protection strategy. Such a preventive strategy must have as its point of departure an increased environmental responsibility for all citizens, so that they might attempt to prevent new problems in all decision-making situations.

The above reasoning becomes clearer if one studies the life cycle of a product. Such studies show that environmental impacts arise during all phases of the life cycle, from mining and processing of raw materials, through the phases of manufacture and consumption, and onwards after the used product is collected for final "destruction". The possibility of minimizing environmental impacts exists along the entire life cycle, and therefore there is a shared responsibility on the part of all actors along and tangential to the life cycle to develop environmentally adapted solutions and procedures. However, it is clear that the manufacturing sector is a key actor since in this sector there are unique possibilities to minimize the environmental burdens of products during their entire lifetime. This presupposes that environmental factors are given importance when raw materials are chosen, products designed and constructed, machines purchased, the conditions and procedures of production decided and the various market relations developed.

However, there are few elements in the traditional command-and-control approach that seek to activate an industrial producer's responsibility in the direction of prevention. Permits for pollution are handed out after a period of negotiation, and evaluation of the permits seldom takes place in a regular and timely manner. Questions concerning solid waste generation and products are seldom covered or even mentioned in permits regulating pollution emissions. The permitting procedure does not contain any inherent dynamic mechanism focusing on the reduced generation of waste and pollution. It has become increasingly apparent that regulatory approaches must be complemented by instruments that work to heighten general environmental awareness, including an increased individual and industrial responsibility to prevent the generation of waste and other byproducts of industrial production.

BASIC PRINCIPLES AND THE TREND OF ENVIRONMENTAL POLICY

In Sweden, the tendency has been to give preventive strategies a more advanced position in environmental policies. The most recent example is the proposal for compiling existing environmental laws into a so-called Environmental Code. This proposal can be seen as illustrating the trend presently dominating Swedish environmental policies, including greater integration of a number of general principles that characterize a preventive strategy. The fundamental principles in the proposed environmental code are the following:

1. Environmental damage should be primarily prevented. Thus the risk of environmental problems should be considered at the outset.
2. Resources extracted from nature must be able to be used and managed in ways that do not damage nature.
3. Those who do, or who may, cause environmental damage or nuisances must pay the cost for preventing or remediating such damage.

These primary principles are supported by the following additional principles:

- *The precautionary principle*, which is expressed in the first primary principle's second sentence. The precautionary principle can be found in the Rio Declaration, Principle 15, where it is defined as follows: "In order to protect the environment, the precautionary approach shall be widely applied by States according to their capabilities. Where there are threats of serious or irreversible damage, lack of full scientific certainty shall not be used as a reason for postponing cost-effective measures to prevent environmental degradation."
- *The polluter pays principle*, which buttresses the third primary principle and is internationally accepted. This principle has been adopted as one of the cornerstones of EU environmental law, and can be found in the Rio Declaration, Principle 16, which states that: "National authorities should endeavour to promote the internalisation of environmental costs and the use of economic instruments, taking into account the approach that the polluter should, in principle, bear the cost of pollution, with due regard to the public interest and without distorting international trade and investment."
- *The principle of best available technology (BAT)* shall, according to the proposed Environmental Code, guide the selection of manufacturing technology, operating procedures and other matters related to production. BAT is considered to correspond to the level of the latest technology development, from technical and economic perspectives, possible to use in any given situation. This principle is also supported in the EU directive concerning Integrated Pollution Prevention and Control, IPPC.
- *The principle of substitution* is, according to the proposal, to be used for products, raw materials, and chemical substances in a way similar to how the principle of best available technology is used with regard to manufacturing and technical equipment. Both principles state that less environmentally dangerous alternatives are to be chosen to replace existing procedures, materials etc. when this is possible.[1]
- *The principle of closed-loop material paths* states that everything extracted from nature in a sustainable fashion must be able to be used, reused, recycled and finally stored or dealt with in ways leading to the least amount of resource use and environmental

damage. According to the proposed new Environmental Code, with regard to goods, this principle will mean that manufactured products must be so environmentally adapted that they can, as far as possible, be reused or the material in them recycled.

All of these principles, some of which are supported through a number of international agreements, have an inherent dynamism that should move developments toward continuous improvements for the environment, including cleaner production. To strengthen and clearly emphasize the highest priority accorded to the preventive strategy with regard to waste, emissions and products, existing environmental policy aims to convert the primary principles into general recommendations, environmental fees and taxes, as well as other general instruments. When compared to past efforts to regulate in detail, these principles and the general instruments ensuing from them afford a greater potential to stimulate innovations and inject continuous dynamism into the area of environmental protection.

The introduction of *extended producer responsibility* can be seen as an illustration of this. By imposing a general physical and economic responsibility on producers, an incentive is created for environmentally adapted product development as well as market-based solutions concerning the management of waste. Through voluntary agreements and/or legally binding regulations, producer responsibility is successively being introduced in Sweden for packaging, tyres, cars, electronic devices, construction materials etc.

Therefore, extended producer responsibility is one of the strategies supporting the eco-cycle principle. By assigning the costs for the final handling of goods to the producer, a result which many economists have considered desirable for quite some time is achieved, namely the integration of environmental costs into the price of goods. In this way, extended producer responsibility is also a measure falling within the polluter pays principle. Based on this principle various environmental fees and taxes are also introduced, which, through penalizing undesirable behaviour, technology and raw materials, also support the use of the precautionary principle, the principle of substitution and the principle of best available technology.

This trend in the use of environmental steering instruments, which the above discussion uncovers, illustrates the following changes: from detailed regulations to general steering mechanisms; from administrative instruments towards economic instruments; from isolated static instruments towards a mixture of market-based instruments with dynamic effects; and from instruments that only regulate production to those that influence the environmental characteristics of products.

THE DEVELOPMENT OF AN INDUSTRIAL ENVIRONMENTAL POLICY

As discussed above, Swedish environmental policies have increasingly been based on a number of general basic principles. The development of strategies and instruments is an ongoing process, with the goal of translating these principles into concrete practice. Environmental policies are therefore exposing manufacturing and service companies to growing demands for the continuous development of new and more environmentally adapted methods and products. Obviously, this places companies and organizations with a passive attitude towards environmental questions in a disadvantageous position.

At the same time as environmental policy has changed character, there has been a radical market improvement for environmentally conscientious firms and "green" goods. Today, the demand for industrial environmental responsibility, previously advanced only by radical environmentalists, is on everyone's agenda. In addition to consumers there are a number of other actors on the market who have legitimate reasons for being increasingly interested in how environmental initiatives are carried out in a firm. Such actors include shareholders, financial institutions, investors (especially those representing so-called green funds), insurance companies and potential purchasers. It is clear that a company with a passive environmental profile can quickly lose in this market, which bases its judgement on completely different criteria to mere compliance to emission permits.

It should be in the interests of industry to develop its own environmental policies with the goal of finding profitable and environmentally sustainable solutions to various environmental problems, instead of adopting a wait-and-see approach to the demands of the market and the requirements of the environmental authorities. The alternative is loss of control in decisions about environmental protection and diminished credibility in a market that is placing increasing value on proactive environmental initiatives and profiles. Powerful national and international organizations representing the interests of business and industry have advocated that all companies and organizations ought to create their own environmental policies. Normative principles and programmes for this purpose have been developed (for example, the ICC *Business Charter for Sustainable Development—Principles for Environmental Management* 1991).

In summary, pressure from regulatory agencies, markets and industrial organizations has led to the conviction on the part of a number of business leaders that it is important to establish environmental management as a priority part of leadership in business. This is still primarily a phenomenon observed in larger companies. However, there are some tendencies toward diffusion to smaller firms, especially the suppliers to larger firms.

A survey conducted at Lund University has shown that only a small percentage of Swedish manufacturing industry can demonstrate environmental efforts based on holistic views, preparedness for the future, responsibility and preventive measures in line with the principles advanced by the regulatory authorities, the market and industrial organizations. However, as many as 25% of the total number of employees in the manufacturing sector are employed in these companies (Arnfalk and Thidell 1992).

An important observation derived from this survey is that the most important factor for environmentally progressive firms is the pressure applied by the regulatory authorities. In firms with over 100 employees, between 70 and 90% of business leaders stated that the requirements of the authorities are the main reason for their firms conducting environmental initiatives. There is no significant difference between large and small firms in this regard, except for the very smallest firms with fewer than 100 employees. Most of these have no environment protection measures, strategies or policies and the pressure from the environmental authorities is not felt to be especially strong or of any particular interest.

ENVIRONMENTAL MANAGEMENT TOOLS PLACED IN A SYSTEM

Regardless of the reasons, it is a fact that a small but growing number of organizations have begun to establish internal environmental policies. The realization of these policies requires an appropriate organizational structure and different types of activities and tools. There is a parallel here to regulatory agencies, which usually have a clear organization and use administrative, economic and information-based steering instruments to achieve their policy goals.

However, when compared to environmental policy in society in general, internal company environmental policy has a short history. The earliest policy statements were formulated during the latter half of the 1980s. Therefore, our knowledge of how to realize proactive and dynamic environmental protection activities in companies is rather limited.

The different forms of environmental management system tools have also been used sparingly by industry. Some firms conduct environmental audits to assess the results of environmental policy. Some firms also have used life cycle assessments in an attempt to determine the environmental impact of a good "from cradle to grave". Other firms have carried out waste minimization opportunity audits in an effort continuously to identify and analyse measures that can contribute to a diminished generation of solid waste and pollution.

Various forms of risk assessment have a somewhat longer history. The tools and strategies of risk assessment have often been used within the chemical industry to diminish the risks of accidents and exceeding emission requirements, among other things. Environmental impact assessments, when used in conjunction with new construction or the expansion of production facilities, should also be included in the category of environmental management tools.

The increasing realization that purchasing can be the path through which a production facility admits future environmental problems has led to the frequent inclusion of environmentally adapted purchasing in company environmental policy. To ensure the use of purchasing as part of an environmental strategy, additional activities and tools are required. In some cases, forms of environmental labelling and so-called sunset lists are used when establishing environmentally related product specifications and demands.

A number of different types of industrial environmental accounting systems have begun to be developed in an effort to ensure that the organization deals with financial matters in a way that is consistent with the overriding environmental policy. Several tools come within the framework of industrial environmental accounting systems, such as environmentally related investment and profitability analysis, as well as green book-keeping and reporting. Common to all systems of accounting are a number of key indicators and, analogous to traditional financial accounting, green key indicators are also being developed to cover environmental issues.

As more activities and tools are used of the types mentioned above, a number of practitioners have come to understand their possibilities and limitations. This has led to the realization that a more systematic use of these tools is required within a framework of a well-structured management system. However, developments thus far have been very much focused on single issues, where the possibility of an overview has been lost because too much faith has been placed in the capacity of individual tools to solve most, if not all, environmental problems. Thus interest has come to focus primarily on environmental auditing and life cycle assessment at the expense of other management tools.

Existing experience shows that, for all activities and tools to be effective in the long term, they should be anchored in a system of planning, management and evaluation of environmental protection measures. When developing systems of this type, a fundamental goal must be to achieve the preconditions for sustainable development through preventive environmental protection and the continuous reduction of environmental burdens. The picture is complicated by demands for the introduction of management systems other than those having a direct bearing on the natural environment. How such management can be appropriately integrated or coordinated with the management of, for example, energy, economy, work environment and quality is an important and relevant problem in this field. No obvious answers to these issues can be given at this time.

ENVIRONMENTAL MANAGEMENT SYSTEMS QUALITY CERTIFICATION

In all work with quality improvement a basic principle is to do things right from the beginning. This principle is equally applicable in the area of the environment. Generally, it is more cost-effective and resource saving to prevent the generation of production waste and pollution than to manage and clean the waste afterwards. Furthermore, it is better to design a product with due environmental consideration on the drawing board than to be forced to make later changes to suit an environmentally aware market.

An environmental consultant with some 10 years' experience of environmental auditing in larger European companies is of the opinion that more than 90% of all environmental problems—even those of a technical nature—are caused by organizational shortcomings (Molenkamp 1992). In the draft documents used in the formulation of the new Swedish work environment legislation, it is claimed that problems in the work environment can often be traced to antiquated organizations, undefined areas of responsibility, shortcomings in procedures, and insufficient planning and management of work in both the short and the long term (ASS 1992).

The progression from attacking the symptoms of the problem to quality assured management systems preventing the generation of problems is, therefore, a natural trend towards both improving quality and diminishing the environmental impact of companies. Thus it is understandable that during recent years the development of a number of norms and standards for environmental management systems has, to a large extent, been inspired by corresponding system developments within quality control and assurance.

In a way analogous to quality management and economic management, environmental management stems from management's responsibility to plan, organize and lead a company or production facility towards its goals. This task has been formalized by the British Standards Institute (1992) in the world's first standard for internal company environmental management systems (BS 7750).

A cousin to BS 7750 is the European Union's Eco-management and Audit Scheme (EMAS) (*Official Journal of the Ecnonomic Communities* 1993), which contains norms for how an environmental management system could be constructed. Within the International Organization for Standardization (ISO), work is progressing on the establishment of internationally accepted standards for environmental management systems (ISO 14001 and ISO 14004) and a number of environmental management tools, e.g. environmental auditing and life cycle assessment (ISO 14000 series).

The increasing environmental demands made within the market, together with the above-mentioned developments in environmental management systems and tools, are paving the way for the establishment of a firm's environmental relations as an important factor. In the future it ought to be completely natural for interested parties to evaluate the environmental management and protection activities of a firm, together with its level of quality and finances, as the basis for determining market position and potential. A certificate of some sort, issued by an independent party, verifying that a firm's environmental work reaches a recognized standard, could come to be an indispensable document for companies in all future relations with customers, regulatory authorities, insurance companies, banks, investors and other stakeholders.

KNOWLEDGE OF VARIOUS ENVIRONMENTAL MANAGEMENT STANDARDS

During summer 1994 a survey was conducted among 500 randomly chosen Swedish manufacturing companies to determine how firm size variation influences relations to and knowledge about environmental management systems. While the average response rate was 50%, there was considerable variation by size. Somewhat over 70% of the large companies (over 1000 employees) responded, while small firms (fewer than 10 employees) had a very low, 10%, response rate. If we consider previous experiences with

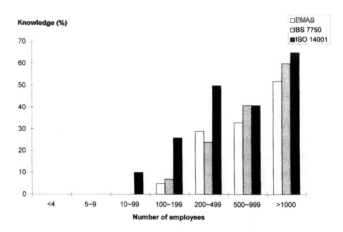

Figure 8.1 Knowledge of environmental management systems by company size

questionnaires concerning environmental awareness in firms, it would appear that the response rate also reflects great differences in knowledge and competence within Swedish industry. However, confirmation of this assumption has not yet been achieved.

As shown in Figure 8.1, knowledge of environmental management standards among the companies that responded appears to be nearly linear and is a function of the size of the company. An exception is the ISO 14000 series about which medium-sized companies appear to have relatively good knowledge.

When answering which standard they find most interesting, 19% of the firms indicated ISO 14001, 8% EMAS and 2% BS 7750. The reasons given for this by medium-sized firms are that ISO 14001 has strong similarities to ISO 9000 standards, together with its international validity.

The survey also uncovered the fact that the interest in being approved or certified according to these standards is presently rather low among Swedish industry. Both EMAS and ISO 14001 received 7% direct interest for participation, while the interest for BS 7750 was only 4%. A greater percentage of respondents, 12% and 16%, expressed doubts or critical views about certification. Thus in a large number of firms that responded, no decision for or against participation had been made.

Willingness to participate is directly proportional to company size. While in the smaller companies (fewer than 100 employees) there is no interest in certification whatsoever, more than 25% of the largest companies (over 1000 employees) have already decided to work towards certification to one or more of the environmental management system standards.

When asked about which factors contribute to a lack of interest in EMAS participation, the primary reason given was "unclear requirements", which naturally reflects a lack of knowledge about EMAS, its purpose and how it is thought to function. Among those who had more knowledge about the system, concerns about costs and bureaucracy seem to be those voiced most often. Another commonly mentioned reason for lack of interest in EMAS is that either the process of ISO 9000 certification has begun or that the firm is already certified to ISO 9000.

Among reasons cited for a firm being positive towards EMAS participation, "pressure from customers" was the most common. A generally expressed environmental interest, as well as ethical and social responsibility, were given relatively frequently for EMAS. The most important results that firms expect from participation include "increased competitiveness" and "a help in marketing". Direct financial effects such as "lower insurance premiums", "raised share price" and "lower environmental protection costs" were considered to be less possible in general. That participation would contribute to "increased worker motivation" and "improved quality" is a rather widespread view.

ADVICE FOR SMALL AND MEDIUM-SIZED COMPANIES

The discussion above illustrates again the observation that environmental competence in small and medium-sized companies is low. This is also the case with regard to preparedness to meet future demands for an approved or certified environmental management system.

The EMAS directive seeks to emphasize the special situation of smaller firms by encouraging member states specifically to support the participation of small and medium-sized firms. In this light, it is surprising that the position paper about the introduction of EMAS into Sweden, written by the Swedish Ministry of the Environment and Natural Resources, primarily concerns itself with the question of which organization should be authorized to register approved facilities (DS 1994). If Swedish business and industry are not stimulated to work more actively with environmental questions, there is a clear risk that this "competent body" will not have too many sites to register.

At this point it might be of interest to note that in the government investigation into environmental taxes or fees on waste (SOU 1994), the establishment of a so-called resource savings committee is proposed. The purpose of this committee is to create educational and informational material, as well as support the use of waste minimization opportunity assessments in small and medium-sized companies through the help of environmental advisers. This investigation also mentions that such assessments could help contribute to the introduction of formalized environmental management systems to ensure a long-term preventive strategy and implementation of such a strategy. It is proposed that the resource savings committee be financed by a special waste fee, which, by itself, will conceivably support the introduction of cleaner production and other waste prevention measures.

CONCLUSIONS

One question remains: if companies in increasing numbers find the new environmental signals from the regulatory authorities, the market and various advisers convincing and therefore implement an environmental management system according to some predetermined norm or standard, will this lead to cleaner production and environmentally adapted products? A possible answer is that the norms offer an important but insufficient *precondition* for such a development. The answer is also partially a function of which norm or standards are adopted.

The EMAS regulation is, without doubt, the most technical and result-oriented norm of those previously mentioned. In addition to following the law, participating companies must commit themselves to achieve "reasonable continuous improvement of environmental performance, with a view to reducing environmental impacts to levels not exceeding those corresponding to economically viable application of best available technology". In Annex 1D, it is stated that a company's environmental policy *shall* be based on a number of basic principles, of which several express an orientation toward cleaner production. The statements in this annex are *binding*.

The British standard BS 7750 is very compatible with EMAS. However, an important difference is that this standard's corresponding description of preventive environmental protection is placed in a *non-binding* appendix (A2). Here, it is stated that a firm *may* state commitments to these principles and measures. The British standard also differs from EMAS by not requiring information on environmental performance to be provided to the public.

Even in the draft proposals for the ISO 14000 series standards there is a requirement for "continuous improvement". The definition of this concept is, however, primarily related to the environmental management system itself and not to physical performance. In addition, as in BS 7750, no public reporting of performance is required. A mandatory policy commitment to "prevention of pollution" is defined in a way that gives no clear direction towards cleaner production.

In general, the trend within international standardization is that directly related environmental questions are placed in non-binding annexes (ISO 1996a) and guidelines (ISO 1996b). Therefore, the standards have a tendency to include lists of building bricks of what an environmental management system should consist of, without ensuring that these bricks are used for cleaner production or in any other environmentally consistent direction. In addition, since no statements are required, the possibility for stakeholders and the general public to obtain information is limited.

It has become increasingly apparent that environmental management system standardization is analogous to the ISO 9000 series standards for quality management. This means that, in the same way that a certified quality system according to ISO 9000 standards ensure an unequivocal quality but not necessarily a high quality, ISO 14001 will ensure a relatively homogeneous construction of certified environmental management systems in industry that will not necessarily lead to a low or diminished environmental impact through cleaner production.

This points to the real risk of a credibility gap, which may very well widen since the standard does not prescribe the possibility of public scrutiny of the results. At the same time, the previous discussion about the similarity between ISO 9000 standards and ISO 14001 may attract firms, especially small and medium-sized firms, to the latter, and possibly lay the foundation for an increased integration between quality and environmental management.

The fact that the result of the survey of Swedish industry shows that many firms see the certification of an environmental management system as a potential advantage in marketing should give us cause to be apprehensive. Considering that the market interest in environmentally produced goods is increasing, it would be most unfortunate if certification is used in such a way that customers and consumers are misled into believing that a certified company is "environmentally friendly" in some objective way. In the long term this could also lead to diminished acceptance of other environmental labelling systems that are of a more objective nature, for example the EU eco-labelling scheme.

At the same time it is clear that the active role that the regulatory authorities play in improving environmental performance in industry will not become significantly less important just because industry has established its own environmental management system. This is equally true for all other actors who are interested in making sure that industry carries out its business and production in an environmentally defensible fashion. For all interested parties, environmental auditing ought to be an attractive tool for ensuring that requirements and expectations are followed. Therefore, it is important that the concept of environmental audit avoids being defined in such a way that excludes any interested party, with the exception of competitors, from gaining insights.

A central issue in this connection, as well as for the credibility of environmental management systems in general, is the question of who or which organization should have the authority to carry out an environmental audit of a facility or company. This so-called client is not given any closer definition in the work leading up to the ISO 14000 series (ISO 1996c, d). Considering that the proposed ISO standard for environmental management systems is rather weak with regard to environmental performance and that public reporting is not prescribed, it is naturally important for the client concept to be given a wide definition.

In summary, it could be said that the threats to the environment are much too serious and their eradication of such public interest that they cannot be left to industry to solve at its own discretion, even if this is done in a systematic way.

NOTE

1. The principle of substitution is incorporated in Swedish environmental policy in the Act on Chemical Products in which it is stated that "Products ought to be avoided that can be replaced by other products which have reduced risks of damage. This requires, among other

things, that companies have well-developed routines in which they consider health and environmental aspects when purchasing chemical products."

REFERENCES

Arnfalk, P. and Thidell, Å. (1992) *Environmental Management in the Swedish Manufacturing Industry: Fact or Fiction?*, Lund University, International Institute Industrial Environmental Economics, Lund, Sweden.

ASS (1992) *Internkontroll av arbetsmiljö*, 6. Stockholm, Sweden.

British Standard Institute (1992) *Specification for Environmental Management Systems*, British Standard BS 7750. London, BSI.

DS (1994) *EMAS—ett system för miljöstyrning och miljörevision. Miljö-och naturresursdepartementet*, 99. Stockhom, Sweden.

ISO (1996a) *Environmental Management Systems Specification with guidance for use: Annex A*, Geneva, International Organization for Standardization.

ISO (1996b) ISO 14004: *Environmental Management Systems General guidelines on the principles, systems and Supporting Techniques*, Geneva, International Organization for Standardization.

ISO (1996c) ISO 14010:*Guidelines for environmental auditing—General principles*, Geneva, International Organization for Standardization.

ISO (1996d) ISO 14011: *Guidelines for environmental auditing—Audit procedures—Auditing of environmental management systems*, Geneva, International Organization for Standardization.

Molenkamp, G.C. (1992) *Experiences with Environmental Auditing in Europe*, The Hague, KPMG Environmental Consulting.

Official Journal of the European Communities (1993) "Council regulations (EEC) No 1836/93 of 29 June 1993. Allowing voluntary participation by companies in the industrial sector in a Community Eco-management and Audit Scheme," L 168, Vol. 36, 10 July.

SOU (1994) *Avfallsfri framtid* (waste-free future). Stockholm, Sweden.

9

Environmental management initiatives in China to promote cleaner production

YA-HUI ZHUANG

INTRODUCTION

The unique features of China's current economic status are: the enterprises' transition to a market mechanism; the coexistence of outdated and advanced technologies and facilities; and the unbalanced development of the coastal and inland regions. The transition to market mechanism brings about a good chance to promote cleaner production (CP) with the aid of economic incentives. The numerous obsolete facilities and outdated technologies offer a wide field for advanced CP technology and management demonstrations, i.e. new policies, regulations, economic incentives etc. These are tested in certain regions, such as the coastal development regions, and, if they are effective, then they are promulgated and applied to the whole country. The unbalanced economic development implies that the coastal cities have greater interest and capability to cooporate with foreign partners to do "build, operate and transfer" (BOT) business.

The obstacles and constraints encountered during the implementation of CP are: lack of financial resources and commercialized (proven) technologies; undeveloped legislation and enforcement systems; and controversial price and tax systems. These problems will be discussed in the following sections.

Environmental Management Systems and Cleaner Production, edited by R. Hillary.
© 1997 John Wiley & Sons Ltd.

INITIATIVES IN CLEANER PRODUCTION

Successful domestic efforts to introduce cleaner processes into individual plants or enterprises have been made in the past decades, e.g. the closed-loop rinsing process in electroplating industry, the recovery of chromium from tannery wastes, the ion-exchange membrane process for the production of caustic soda, and the iron-based catalyst system in the production of butadiene. Since the late 1980s, some joint ventures with foreign companies have led to the import of cleaner technologies and equipment, for example water-based paints, low-CFC or non-CFC refrigerators and new surface-cleaning technology. Even the dry-ice tobacco-expansion process is being transferred to China.

However, all these domestic and cooperative activities constitute only scattered and individual cleaner technologies among a vast ocean of outdated facilities and backward technologies. The outdated facilities and equipment can be traced back to before the Second World War. One can find antiques from the 1930s, obsolete Soviet equipment from the 1950s, domestic facilities from the 1960–70s and more advanced facilities from western countries imported during the 1980–90s. Strictly speaking, there is no systematic cleaner production in any industrial sector in the sense of a "cradle-to-grave" approach.

Since the first two national workshops on cleaner production were held in Xiamen and Shaoxing in 1992, sponsored by the Chinese National Environmental Protection Agency and the United Nations Environment Programme Industry and Environment Programme Activity Centre (UNEP IE/PAC), some CP efforts have been initiated in China. The World Bank has given loans to promote cleaner production demonstration projects in China. IE PAC/UNEP participates in the design and preparation of auditing methodology. The first batch of seven test enterprises have finished the preparatory stage of their demonstration projects. United Nations Industrial Development Organization (UNIDO) also began to be involved in the formulation and implementation of policy measures for ecologically sustainable industrial development (ESID).

EVALUATION OF THE ENVIRONMENTAL MANAGEMENT SYSTEM TO PROMOTE CLEANER PRODUCTION

The existing national environmental management system is oriented toward pollution control and relies on environmental standards and an interim permit system, but does not included dimensions such as raw material consumption and energy consumption. It does not meet the requirement for the implementation of CP from "cradle to grave", because there is:

- no systematic methodology for the implementation of CP;
- no systematic demonstration of CP in any sector of industry;
- no insight into the barriers to the implementation of CP;
- no integrated policy for the implementation of CP.

For example, a primitive enterprise environmental management appraisal system formulated in 1986 for non-ferrous metallurgical companies is shown in Table 9.1. The emission standards do not encourage firms to curtail their discharges and emissions sharply. There should be a step-by-step tightening of emission standards and also be a spatial differentiation of emission standards. Along with the environmental quality standards, there should be regulations on the energy, raw material and water consumption per unit product.

Table 9.2 gives suggestions for the amendment of the Air Act in China with the temporal, spatial and economic differentiation of regulations and standards and their gradual tightening. As the regional economy develops, the environmental quality standards and waste discharge (quantitative) regulations become more stringent.

ECONOMIC TOOLS IN THE PROMOTION OF CLEANER PRODUCTION

Natural resource utilization fee

Crucial issues encountered during the implementation of CP are the irrational prices of fresh water and coal in China. The current prices do not reflect the actual exploitation costs of these natural resources, since the government paid for the complete capital costs of the construction projects for irrigation canals and water supply systems, and did not expect capital payback from the consumers. Consequently, the price of fresh water for industrial use is much lower than the cost of on-site water recycling (Table 9.3).

For this reason, polluters do not have any incentive to recycle their waste water. They prefer to discharge waste water and pay the fines, rather than reuse the waste water or treat it to meet the environmental standards. Experts in China are aware of the irrational price problem; however, price adjustment seems to be a difficult matter and will not be settled in the near future. The simplest way to solve this problem is to set norms for raw material consumption for cleaner production, including norms for fresh water, fuel, air consumption and land use per unit product.

For example, the Beijing Water Company allots water to all communities and enterprises. If consumption exceeds the norm, the excess amount of fresh water consumed is charged at a rate three to five times higher than the normal one. This economic tool has promoted water conservation in China.

Mix of economic tools for energy price adjustment

Coal pricing is a more complicated issue than water. Since cleaner fuels such as natural gas and oil can be purchased at fairly modest prices on the international market, consumers could shift from coal to imported oil or gas. This fuel shift would favour environmental quality improvement and CP, but the government would suffer

Table 9.1 Environmental management appraisal criteria in the non-ferrous industry (issued by the China Non-ferrous Industrial Corporation in 1986)

Categories	Performance	Grade points
I. Overall operation, economic and environmental returns	1. All technical and economic returns reach the norms set by the enterprise.	30
	2. Sulfur recovery rate > 85%; $HCl + Cl_2$ recovery rate > 70%; red clay make-up rate > 25%.	25
	3. Water reuse rate reaches the norm set by the industry.	30
	4. Economic benefits of waste minimization are among the highest in the industry.	15
II. Effluent standards	1. The effluent quality and quantity meet the standards.	35
	2. Dust concentration on spot < 80% of the standard value.	15
	3. Mine ore tailings and slag should not be released into rivers, lakes or farmland.	20
	4. No secondary pollution from tailings, slag and byproducts.	10
	5. No pollution accident.	10
	6. Good pollution-control facilities.	10
	7. Pollution-prevention/control projects accomplished on time.	10
III. Environmental management and monitoring systems	1. *Administrative responsibility:* The company shall have one manager in charge of environmental protection affairs and one member of staff in each workshop for the same purpose.	5
	The management shall regularly review the environmental protection work annually.	5
	The responsibilities and rights of the environmental officers are clearly defined.	5
	2. *Management:* Regulation promulgation and implementation.	10
	Complete monitoring system and regular report.	10
	Publication of annual statistics on environmental protection.	5
	Complete record-keeping since 1980.	5
	3. *Planning:* Long-term and annual planning and implementation.	5
	Annual executive summary report.	5
	Environmental tasks fulfilled as planned.	5

continued overleaf

Table 9.1 (cont.)

Categories	Performance	Grade points
III. Environmental management and monitoring systems (cont.)	4. *Facilities:* Environmental and production facilities are in operation simultaneously. More than 90% of the facilities are in good conditions. Technicians are designated to the maintenance of the environmental facilities.	5 5 5
	5. *Pollution prevention and control:* Adequate funds are available to support PP plans. Measures are available to prevent or control waste formation from the new, expansion or upgrading projects launched since 1980.	10 5
	6. *Public awareness stimulation:* Stimulating the public awareness through all the enterprise's media. Training on environmental protection at all levels. Strict education for new employees.	4 3 3
IV. Greening of the production and living environment	1. Clean facade and cleared roads. 2. Clean and neat production and operation sites. 3. No less than 90% of the "greenable" areas are in green coverage. 4. No "visible" pollution in the living quarters and production areas; the air quality meets the industrial and residential standards.	25 15 30 30

from serious hard-currency loss. The government could not support such a fuel shift, because China has abundant coal reserves but a shortage of oil and gas. The government could rely on multiple economic tools to ameliorate the energy and environment conflict. The energy price system is changing to fix the price of inferior quality coal and allow the prices of superior quality coal to rise. Meanwhile, energy-saving measures and relevant economic incentives are being taken into consideration. A similar norm-allocation approach has been adopted.

Phasing out obsolete energy-wasting facilities and replacing them with modern energy-efficient ones is underway. Construction of highly efficient coal-fired power plants has been speeded up. Newly constructed power plants will be required to have both flue gas desulphurization and dust-removal devices installed. Depreciation of the investment will be included in the utility fee. Joint ventures with foreign companies, either in the design and construction phase or in the operation phase, are the main measures to bring high technology and capital into the cleaner production of the power industry. Construction of municipal coal and gas supply systems is encouraged

Table 9.2 Dimensions to be included for the integration of policies

Types of criteria	Indices
Technical performance criteria	Energy, water and raw material consumption per unit product
Environmental performance criteria	i. Quantities of wastes produced per unit product ii. Quantities of pollutants produced per unit product
Economic performance criteria	i. O & M cost (including waste recycling and treatment cost) per unit product ii. Investment per unit product iii. Payback period iv. Annual revenue
Regional economic priority	Class I: GNP per capita > RMB 8000 Class II: GNP per capita < RMB 8000 but > RMB 4000 Class III: GNP per capita < RMB 4000
Regional vulnerability priority	Class I: high vulnerability, e.g. acidic soil areas or pristine areas Class II: moderate vulnerability, e.g. neutral soil areas or residential districts Class III: low vulnerability, e.g. basic soil areas or industrial districts
Grace period priority	Class I: immediate implementation Class II: short grace period Class III: long grace period

Table 9.3 The irrational price of fresh water as compared with the cost of recycling

Price of tap water	Waste water discharge fee	Cost of waste water recycling	Cost of end-of-pipe treatment
RMB 0.50/ton	RMB 0.05–0.10 /ton	RMB 0.70–1.00 /ton	RMB 1.50–2.50 /ton

by granting low-interest loans. All these tools became effective only after economic reform, i.e. the transition to the market mechanism.

Deposit system for returnable waste collection

The deposit system has been proved to be an effective economic instrument for the collection of returnable beverage bottles. The deposit follows the bottle from the brewery and retailer to the consumer. It is then returned through each stage to the consumer in the bottle collection process. Extending this practice to other wastes is recommended, such as fluorescent lamps, batteries, accumulators and other containers and packaging.

PRIORITY SETTING FOR CLEANER PRODUCTION OPTIONS

Because CP options may require considerable investment, its careful implementation is necessary. Some of the CP options, such as good housekeeping and minor modification in production processes, are of low cost or even cost free. Their implementation usually encounters no or few obstacles. However, on-site and off-site recycling or recovery requires both technology input and investment, and the payback period could be long. In these cases, the obstacles are larger, and demonstration projects are necessary. With sophisticated cleaner processes, it is not only the cost and technology factors but also the market risk factor which should be considered.

Table 9.4 shows the various phases of CP in the tanning industry. It is easy to understand that the adoption of new processes and waste recycling takes a longer period and higher investment (Zhuang 1992) and that the risk in finding a market is higher. For instance, there are many patented chromium-free processes (Nagarur

Table 9.4 Effective measures for implementation of various phases of CP in small and medium-sized tanneries

Phases of CP	Cost	Time period	Barriers or constraints	Incentives or measures	Implementation priority	CP priority
Good housekeeping	None	Short	Responsibility	Training; reward	1	5
Process modification	Low	Short	No initiatives	training; reward; funding	3	3
Waste water recycling	Medium to high	Medium	Lack of capital and technology	funding; training	4	4
Recovery of chromium from leather shavings	Medium	Medium	Lack of capital and technology; market risk	loans; training; series of products	6	7
Utilization of leather trimmings	Low	Short	Market risk; training		5	6
Chromium-free tanning processes	High	Long	Small market; investment risk	Eco-labelling	7	1
Shaving incineration	High	Medium	Capital; management; environmental risk		8	8
Alternative artificial leather	High	Medium	Capital	Eco-labelling	2	2

1992). To a layperson these sound valid and seem to be able to replace the chromium process. To the experts, the leather produced by the chromium-free processes gives in some cases a softer feel to the touch, in others a harder feel. This implies an uncertainty in marketing. Governmental subsidy and eco-labelling might help to overcome the barriers to chromium-free leather.

ECOLOGICAL PLANNING AS AN EFFECTIVE TOOL FOR TOWNSHIP ENTERPRISE REFORM

Township enterprise reform is a unique issue in China's CP management. In order to prevent indiscriminate and wasteful exploitation of natural resources and pollution transfer, the government should foster economic-cum-environmental planning of towns and villages and promulgate guidelines to reform their activities. On the request of the Chinese EPA and the local governments of Tianjin, Ma'anshan and Dafeng, our Research Centre has been involved in ecological planning of cities and towns (Yang et al. 1989; Yang et al. 1992). An index system has been developed for the assessment and screening of industrial activities in these cities and towns.

The term "ecological planning" can be defined as the planning of multistage utilization of natural resources such that the wastes from the preceding stage of production can be used as the raw materials in the successive stage, just as in a natural ecosystem. In natural ecosystems there is virtually no waste. All the excrement, faeces, litter etc. can be utilized by other biological species as food. This is the result of lengthy evolution. The final goal of cleaner production, after a lengthy evolution of processes, would be an ideal integrated production system, in which each waste flow can be transformed into valuable byproducts with very low energy consumption.

The index system is shown in Table 9.5. The outcome of this planning in Dafeng county, for instance, is a series of production lines or "trees". Since this county does not have mineral resources, polluting and energy-intensive industries such as metallurgy and power were ruled out. The initial resources are cotton, barley, corn, soya bean, mulberry, rapeseed, reed and coastal wild land.

For instance, the mulberry gives silk, fabrics and clothes as its main products. Its byproduct is pupa, which can be used as the raw material for many valuable biochemical products such as protein, nucleic acid and cosmetic.

The planning of the coastal wild land led to the establishment of a David's deer preserve and a crane habitat, which will give impetus to the local tourism business.

Barley is treated under suitable temperature and moisture conditions to produce sprouts that are used in a brewery to produce beer. The distillers' grain is used as the feed for hogs. The waste water from the brewery is contained in an oxidation fishpond system, where organic matter in the biochemical oxidation demand (BOD) is assimilated by algae. The algae serve as feed for the fish, while the fish serves as the prey of ducks. The hogs, ducks and fish are slaughtered, frozen or preserved in cans,

Table 9.5 A hierarchical system for evaluation and decision support for township regulation

First-level indices	Second-level indices	Third-level indices
Efficiency	Resource utilization efficiency	Raw material consumption; per unit output; waste utilization rate; land-use yield
	Energy transformation efficiency	Fossil fuel and renewable fuel utilization efficiency
	Technology transfer	
	Production efficiency	GNP per capita
	Capital turnover rate	Revenue per unit investment (RMB 10 000); increment in value during byproduct processing
Benefits	Economic benefits	Revenue change rate; increase rate of GNP per capita; reproduction expansion capability
	Life quality improvement	Commodity supply; purchasing power; social security; traffic and communication; medical and public service; education; social welfare insurance; cultural life
	Environmental quality improvement	Air, water, soil; landscape; noise; forest coverage
Maturity	Recycling function	Stagnancy and exhaustibility; recycling rate of wastes
	Synergetic function	Network diversity; openness towards outside world; system rationality
	Vitality towards sustained self-dependence	Attraction to capital investment; attraction to qualified personnel; decision and management ability; ecological consciousness; reliability of external support

Source: Yang et al. 1989

while the bones and hides are used as the raw material for food additives, hair brushes and leather. The feathers are used for sportswear.

This kind of planning helps the local government to make decisions on the reform of township enterprises. The planning is based on the natural resources of China. Low-pollution processing industries with low energy consumption are planned, while high-pollution and energy-intensive industries are banned.

CONCLUSIONS

The transition to the use of market mechanisms brings about a chance to promote CP through economic incentives such as discharge fees, consumption norms and deposit systems. A feasible tactic for promoting CP would consist of:

- a quick start to improve housekeeping and to make minor modifications to production processes through simple auditing with direct economic benefits;
- a medium-term project of recycling and retrofitting demonstration projects, together with the formulation of appropriate criteria and standards for CP implementation;
- a long-term programme to establish corporate enterprises to build, operate and transfer (BOT) those capital-intensive CP facilities and high technologies.

REFERENCES

1. Zhuang, Y.H. (1992) "Profitability of protein recovery from leather shavings with high-level chromium content", *Proceedings of the International Seminar on the Profitability of Clean Technology in the Leather Tanning Industries*, Samutprakarn, Thailand, October 20–21.
2. Nagarur, N.N. (1992) "Prospects of toxic wastes minimisation through process modification", *Proceedings of the International Seminar on the Profitability of Clean Technology in the Leather Tanning Industries*, Samutprakarn, Thailand, October 20–21.
3. Yang, B.J., Wang, R.S., Lu, Y.L., Hu, X.L., Deng, X.F. and Lu, L. (1989) "An intelligent decision support system for urban ecological regulation", in Lui, J.G. (ed.) *Human Ecology in China*, Beijing, China Science and Technology Press.
4. Yang, B.J., Hu, X.L. and Zhong, Y.G. (1992) "Application of Decision Support System to Urban Ecological Regulation", *IIASA Workshop on Intelligent Decision Support Systems*, Kutzively, Ukraine, October 1–8.

10
Hong Kong's experience with cleaner processes

EDWIN KON-HUNG LUI

GEOGRAPHICAL AND POPULATION DATA OF HONG KONG

Hong Kong is a small territory situated on the South China coast. It has a total area of just over 1000 sq. km. and had a population of around 6.2 million in mid-1995. Hong Kong can broadly be divided into three regions, namely Hong Kong Island, Kowloon Peninsula, and the New Territories and outlying islands. As the mainland and islands of Hong Kong are a partly submerged old mountain range there are many steep and hilly areas and the overall usable areas for residential, industrial or commercial purposes are limited. This leads to a population density of over 26 000 people per sq. km. in some of its urban areas.

HONG KONG'S INDUSTRY AND CLEANER PROCESSES IN GENERAL

Hong Kong is one of the world's most successful communities and has sustained a steady growth in its economy. Within the span of two generations, this economic growth has significantly improved the living standards of its people. The average annual real gross domestic product (GDP) growth was 6.5% during the last decade. Hong Kong's per capita GDP is among the highest in Asia and is higher than that in some European countries.

Manufacturing industry is still one of the pillars of the Hong Kong economy, although less polluting economic activities, such as financial services, the container port and tourism, are new engines for growth in the economy.

Environmental Management Systems and Cleaner Production, edited by R. Hillary.
© 1997 John Wiley & Sons Ltd.

There are quite a number of industrial sectors in Hong Kong; the four major ones, in terms of total domestic exports in descending order, are clothing and textiles, electronics, watches and clocks, and plastics (Chan, S.T. 1994). In each of these sectors, operators use technologies of various complexities. Due to favourable labour costs and land supplies, some of Hong Kong's industry has been drawn into China.

Even though there are two purpose-built industrial estates and a third industrial estate is under construction, Hong Kong's industry mainly comprises small factories housed in multistorey industrial buildings in the urban areas. These high-rise industrial buildings are not normally equipped with any communal waste treatment facilities and effluent discharges from the factories often go directly to the government sewage-disposal works without pretreatment. Each of these "flatted" factories only occupies part of a multistorey building; it is usually a small business and has crowded working conditions. The majority of Hong Kong's industrial establishments employ less than 50 workers.

All industrial establishments, irrespective of their employee numbers and nature of business, have to comply with the legislative requirements for the protection of air, noise, waste and water specified by the relevant environmental ordinances and their subsidiary regulations. Most waste producers in Hong Kong rely on end-of-pipe treatment of effluents and emissions to meet the regulatory requirements; in other words, clean technology has not been extensively applied in the territory. Many manufacturers are inclined to stick to their adopted production procedures which include end-of-pipe technologies.

Manufacturers consider that:

- process changes will usually involve huge initial capital investment;
- production will have to be suspended to establish the optimized process conditions for the new procedures;
- waste disposal costs in Hong Kong are relatively cheap; at the time of writing this chapter disposal of waste at landfills is free.

Nevertheless, some establishments have switched to cleaner production processes in order to cope with the legislative emission and discharge standards.

To ensure that waste producers undertake thoughtful consideration and in-depth analysis of the concept of cleaner production, the Hong Kong government has devised several schemes and measures.

CHEMICAL WASTE MANAGEMENT IN HONG KONG

Measures to stimulate the adoption of cleaner processes have been incorporated into the management of chemical waste in Hong Kong. The term "chemical waste" refers to wastes which exhibit chemically hazardous properties. All industrial wastes that exhibit hazardous properties are now generally under regulatory control. In the past, waste producers who were relatively small in terms of space or number of workers and whose establishments could not afford to employ professionals to treat their wastes simply discharged their chemical wastes into the drain. This malpractice posed serious health threats to factory workers and sewer workers. More importantly, the wastes ended up in the sea, leading to poor water quality and polluting the marine environment.

In view of this, the Environmental Protection Department appointed a contractor to design, build and operate an integrated chemical waste treatment centre (CWTC) to provide treatment services for most types of chemical wastes and to serve as a major, proper disposal route for chemical wastes from industry. The contractor of the facility also provides door-to-door waste collection services. Parallel to the commissioning of the facility, the department has also implemented "cradle-to-grave" control of chemical waste through the enactment of the Waste Disposal (Chemical Waste) (General) Regulation which became fully effective from 3 May 1993 (Hong Kong Environmental Protection Department 1994).

In the initial stage, collection and treatment services were free of charge. The government has considered several alternatives to recover the operating costs of the CWTC. After comprehensive consultation with the public and private sectors, it has adopted a direct charging scheme which requires waste producers to pay for the collection and disposal costs according to the type and quantity of wastes. The scheme has been implemented since March 1995 and charges will gradually increase to achieve full recovery of the operating cost for each waste stream.

It is considered that a phased charging scheme will have minimal impact or burden on industry. At the same time, paying for waste disposal will send a clear message to industry about waste minimization and the "polluter pays principle". This should be regarded as one of the steps to attract waste producers' attention to cleaner processes in Hong Kong.

Disposal costs will ensure that waste producers seriously consider the various options available for waste minimization and cleaner processes. It is envisaged that small to medium-sized factories will primarily focus on simple and low-cost procedural changes, housekeeping and/or substituting environmentally friendly process reagents for existing ones. Significant process changes or modification which involve high investment will depend on the break-even and payback periods.

To summarize, the Environmental Protection Department's overall approach to chemical waste management comprises the following:

- the provision of an environmentally sound disposal outlet through the construction of an integrated chemical waste treatment facility;
- the implementation of legislative control on storage, collection, transportation, treatment and disposal of chemical waste;
- the promotion of waste reduction and minimization at source through the implementation of a phased direct charging scheme.

In the next few years it is expected that industrialists will direct more attention and effort to cleaner processes (Chan, Y.L. 1994). Table 10.1 highlights some examples of cleaner processes currently practised by Hong Kong's manufacturers.

Table 10.1 Some examples of applications of cleaner processes in Hong Kong

Industrial sectors	Clean technology applied
Bleaching and dyeing	Use of low-impact dyes which are more efficient in fixing to cloth compared with traditional dyes Regeneration of caustic from spent alkaline mercerization solution through filtering and vacuum evaporation of water Reuse of spent alkali solution generated from scouring
Construction industry	Use of recyclable materials (such as steel) instead of hardwood as hoarding and formwork at construction sites Use of pulverized fly ash as a substitute for cement in concrete work
Electroplating of nickel, copper, silver, gold	Installation of metal recovery systems to recover toxic metals for reuse, e.g. nickel recovery from its sulphate solution Installation of drag-out tanks to collect the first rinse waste water and to reuse as make-up water for the plating bath Reuse of second rinse waste water
Film production and photoprocessing	Substitution of toxic cyanide-containing reagents by non-toxic proprietary ones, cyanide and other toxic substances in the waste water eliminated Use of automatic photoprocessing machine with built-in unit for silver recovery
Land transport	Use of unleaded petrol and asbestos-free brake linings Reuse of brake shoes through recycling
Motor servicing	Recovery of spent halogenated solvents used for the cleaning of spare parts by *in situ* distillation
Paper recycling	Recovery of fibres from process water
Plastic manufacturing	Installation of dioctyl phthalate (DOP) recovery system, eliminating white plasticizer mist and unpleasant odour and allowing the recovery of the costly DOP for reuse
Printed circuit manufacturing	Recovery of spent etching solution for reuse and production of pure copper sulphate for sale Incorporation of counter-current rinsing techniques for reducing water consumption and effluent discharge Reuse of spent trichloroethylene to wash printed circuit boards Replacement of halogenated solvent degreasing reagents with less polluting aqueous proprietary chemicals

OTHER MEASURES FOR CLEANER PROCESSES IN HONG KONG

The Hong Kong government has utilized various channels to promote to the public and industrialists that the application of cleaner processes not only helps to protect the environment, but also leads to economic benefits through saving of valuable resources and reducing the costs of waste treatment and management (Planning, Environment and Lands Branch 1994). The following are some examples of activities launched by the government with respect to cleaner processes:

1. A consultancy study on waste reduction is being conducted which aims to devise practical measures to further waste reduction in Hong Kong. The study was started in February 1994 and Phase I was completed in late 1995. The government is now considering the consultant's recommendations and will work out a detailed waste reduction plan for Hong Kong.
2. The Environmental Protection Department started a consultancy study on life cycle assessment (LCA) and eco-labelling in February 1996. The study focuses on examining the applicability of eco-labelling schemes and green purchasing in Hong Kong.
3. A direct charging scheme for disposal of solid wastes to landfills will be implemented and the relevant legislation has already been enacted. The scheme is necessary to encourage waste producers to practise waste minimization, recycling and waste segregation at source, and to reserve valuable landfill space for wastes which really need to be disposed of at landfills.
4. The Hong Kong Industry Department has conducted a consultancy study on helping industry with environmental affairs, set up an environmental hotline and a directory for use by local industry, and prepared handy reference books on waste management for use by three specific industries: electroplating, bleaching and dyeing, and printed circuit board manufacturing.

CONCLUSION

The Hong Kong government has incorporated suitable measures and incentives into its overall waste management plan to encourage waste producers to take steps and initiatives in waste reduction and cleaner processes. Industry is becoming more aware that the concept of pollution prevention and cleaner processes has replaced the traditional end-of-pipe treatment approach as a means to deal with environmental pollution. The government and manufacturers are actively working towards the adoption of cleaner processes in Hong Kong as these processes are cost-effective and can bring about economic benefits.

ACKNOWLEDGEMENTS

The author wishes to express his sincere thanks to colleagues from the Waste and Water Division and the Local Control Division of the Hong Kong Environmental Protection Department for their contribution in providing information on examples of cleaner processes practised in Hong Kong.

The information contained in this paper may be freely used. Acknowledgement of the source must be made and the Director of Environmental Protection notified.

REFERENCES

Chan, S.T. and Wong, C.F. (1994) "Hong Kong environmental legislations and institutional arrangements on waste and water management in relation to green productivity", presented at workshop on *Green Productivity in Small and Medium Enterprises*, organized by the Asian Productivity Organization, Hong Kong, May.

Chan, Y.L. and Leung, R.Y.C. (1994) "An overview of clean technologies in Hong Kong industry", presented at Polmet 94, co-organized by Hong Kong Institute of Engineers and Hong Kong Environmental Protection Department, Beijing, November.

Hong Kong Environmental Protection Department (1994) *Environment Hong Kong 1994*.

Planning, Environment and Lands Branch, Government Secretariat, Hong Kong Government (1994) *A Green Challenge for the Community, 2nd Review of 1989 White Paper: Pollution in Hong Kong—A Time to Act*, Hong Kong Printing Department, Hong Kong.

11
Linking cleaner production and ISO 14001 environmental management systems in New Zealand[1]

MARJE RUSS

NEW ZEALAND'S BUSINESS ENVIRONMENT

During the 1980s the New Zealand economy underwent substantial reform. New Zealand's business environment is now a highly competitive, unprotected and export-focused market. It is also an environment with a very high proportion of small to medium-sized enterprises. More than 80% of New Zealand's 200 000 businesses have five or fewer employees and only 1.5% have 50 or more employees. In this environment, New Zealand businesses have become very customer focused. To establish and maintain a place in the international market, companies have had to be able to meet, and to show that they can meet, their customers' requirements. Increasingly this includes requirements to demonstrate sound environmental performance.

Quality management systems, especially independently certified systems, have become an important factor in the success of many New Zealand businesses. Such systems provide evidence that companies can meet their customers' quality requirements. Uptake of ISO 9000 standards by New Zealand businesses has been higher than in many other countries. More than 1000 New Zealand companies are now registered to ISO 9000 and around 2000 more are in the process of implementing

---

Environmental Management Systems and Cleaner Production, edited by R. Hillary.
© 1997 John Wiley & Sons Ltd.

ISO 9000 systems. New Zealand has also developed Q-Base,[2] a quality system especially designed for small businesses. With the high proportion of small businesses in New Zealand, it is not surprising that nearly 800 businesses are already participating in Q-Base.

New Zealand businesses have appreciated the advantages of having a disciplined approach to quality, with independent certification as evidence that they meet international standards. They are well placed to extend this approach and gain similar advantages in the area of environmental management.

NEW ZEALAND'S REGULATORY ENVIRONMENT

New Zealand's environmental law has also undergone significant change. In 1991, a major programme to reform resource management law concluded with the introduction of the Resource Management Act. The Resource Management Act 1991 is a very important piece of legislation and its influence, as a model for environmental regulation, extends well beyond New Zealand.

One of the reasons the Act is so significant is its purpose, which is:

> *to promote sustainable management of natural and physical resources* (land, air, water, soil, minerals, energy, plants and animals, and structures).

Sustainable management is a New Zealand interpretation of the internationally recognized term "sustainable development". "Sustainable management" is defined in the Act as:

> Managing natural and physical resources so that people can provide for their social, economic and cultural well being, health and safety while:
> - sustaining the potential of resources to meet future needs
> - safeguarding the life supporting capacity of air, water, soil and ecosystems
> - avoiding, remedying or mitigating adverse effects on the environment.

This definition is important as it establishes an effects-based approach to environmental management. In the past, planning and environmental regulation in New Zealand controlled activities. Under the Resource Management Act, it is the environmental *effects* of activities that are controlled, not the *activities* themselves. The Act places a general duty on everyone to avoid, remedy or mitigate the adverse environmental effects of their activities.

This effects-based approach means that regulators' policies, plans and rules focus on environmental outcomes rather than on prescribing how activities should be carried out. In turn, businesses need to manage and regulate their own activities, and the effects of those activities, to meet the environmental outcomes specified in policies, plans or individual resource consents. A similar philosophy, requiring businesses to manage their own performance in order to meet broad statutory objectives, underpins New Zealand's Health and Safety in Employment Act 1992 and law reform currently underway for hazardous substances and new organisms.

ENVIRONMENTAL MANAGEMENT SYSTEMS IN NEW ZEALAND

Businesses in New Zealand have been developing approaches to environmental management for a number of years. Since the late 1980s, environmental audits have become a common tool for environmental management. Much auditing activity has been prompted by businesses wanting to understand their responsibilities and status under the new legislation. Environmental auditing led to some businesses developing action plans or rudimentary environmental management system (EMS). This activity was significantly assisted by the development and publication in 1992 of BS 7750, the British Standard for EMS, and publication of the draft international standard, DIS ISO 14001, as an interim standard in New Zealand in November 1995.

BS 7750 has been used by companies in New Zealand as a model for EMS from the start of 1992. Businesses familiar with the ISO 9000 approach to quality management have responded quickly and positively to applying the same kind of principles and approach to environmental management. Independent certification of EMS has been available in New Zealand since the beginning of 1994 and a number of New Zealand companies are formally seeking certification. New Zealand also provided significant input to the development of the international standard ISO 14001 and business interest in the new standard is high.

In the early 1980s, the chemical industry in Canada developed a code for environmental, health and safety management called Responsible Care. This programme was formally introduced to New Zealand by the New Zealand Chemical Industry Council (NZCIC) in March 1991. The code and associated handbooks have also provided guidance to New Zealand businesses implementing environmental management. All members of the NZCIC are required to adopt the Responsible Care code.

CLEANER PRODUCTION IN NEW ZEALAND

New Zealand businesses were involved in projects to minimize wastes and reduce energy use well before the term "cleaner production" (CP) became common. CP has, however, provided an umbrella to draw together information and examples of the links between good environmental management and enhanced economic performance.

CP has been promoted in New Zealand to help achieve sustainable management of natural and physical resources. National CP initiatives in New Zealand have concentrated on providing case-study information to demonstrate the benefits which New Zealand companies have obtained by applying the principles of CP. The Ministry for the

Environment has published this information in two publications (1993 and undated). Some of the information has been generated through a programme of demonstration projects. This programme involved guidelines being produced on CP methodology and training for participating companies on the principles and methodology for CP.

CP initiatives have also been taken at a regional and local level. Northland Regional Council has produced a set of booklets on CP. Booklets have been published covering industries such as panel beating, commercial printers and spray painters. Wellington City Council is working with a number of businesses in the city as part of its Workplace Pride programme and Dunedin City Council has a Green Business Challenge programme. Other similar initiatives are underway in other areas.

Published results from companies which have been involved in the CP demonstration projects have shown that what is good for the environment can also be very good for the bottom line. The desire to achieve greater efficiencies and cost savings has been a key driving factor for most of the participating companies and the outcomes are very impressive. Projects to reduce use of raw materials, water or energy have delivered savings ranging from 12–80% of annual costs. Projects to reduce waste disposal costs have saved between 50 and 100% of annual costs and, in cases where reuse has been involved, have produced extra income. Payback times are also good, with some projects paying for themselves in a matter of days or weeks.

LINKS BETWEEN ENVIRONMENTAL MANAGEMENT SYSTEMS AND CLEANER PRODUCTION

There are clear and strong links between EMS and CP. These can be shown by comparing the elements involved in an EMS with the steps in the methodology developed for CP demonstration projects in New Zealand (Ministry for the Environment, undated).

ISO 14001 sets out the requirements for an EMS. There are five main elements that make up the ISO 14001 EMS model. Within three of these main elements, there are more specific elements (see Figure 11.1).

There are six main steps in the CP method. Most of these steps consist of a number of sub-steps (for details see Table 11.1). The six steps are:

- plan and organize;
- review data and site(s);
- identify CP options;
- evaluate the options;
- choose your project(s);
- implement and monitor initiatives.

The links between the elements in an ISO 14001 EMS and the steps in the CP methodology are outlined below. The ISO 14001 elements are defined first, followed by the equivalent CP steps. A summary of these links is presented in Table 11.1.

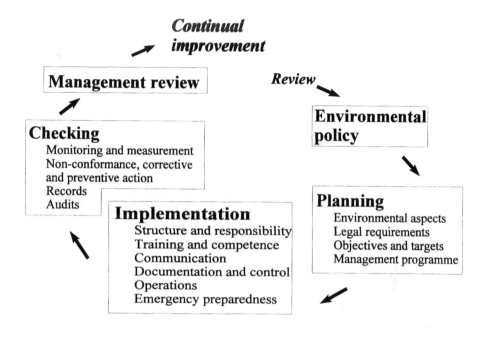

Figure 11.1 The ISO 14001 environmental management system elements

Environmental policy

ISO 14001 requires an organization to set an environmental policy which is appropriate to the environmental impacts of its activities, products and services. The policy must make commitments to continual improvement, pollution prevention and compliance with the law. It must also provide a framework for setting and reviewing environmental objectives. CP equivalents are steps to obtain management commitment (which requires a policy) and to set broad project objectives.

Environmental aspects

ISO 14001 requires an organization to have procedures to identify aspects of its activities, products and services that may interact with the environment and to assess how significant those impacts are. This information must be kept up to date and be considered when environmental objectives and targets are set. CP equivalents are

Table 11.1 Links between cleaner production and ISO 14001 environmental management systems

Cleaner production steps	Policy	Aspects	Legal	Objectives and targets	Programme	Structure	Training	Communication	Documents and control	Operational control	Emergency	Monitoring	Non-conformance	Records	Audits	Reviews
PLAN AND ORGANIZE																
Obtain management commitment	✓															
Select cleaner production team						✓	✓									
Set project objectives and timeframes				✓	✓	✓	✓									
Identify potential difficulties and solutions		✓											✓			
REVIEW DATA AND SITES																
Collect initial data		✓	✓													
Inspect the site		✓	✓													
Identify potential areas for improvement		✓	✓		✓	✓										
Establish support teams						✓	✓									
Collect detailed data		✓	✓													
IDENTIFY CLEANER PRODUCTION OPTIONS																
List possible ways to improve in each area		✓	✓	✓	✓											
Hold a team meeting							✓	✓								
EVALUATE THE OPTIONS																
Technical and practical evaluation				✓												
Environmental evaluation				✓												
Economic evaluation				✓												
Draw information together				✓	✓											
CHOOSE YOUR PROJECTS				✓	✓											
IMPLEMENT YOUR PROJECTS																
Prepare your project					✓				✓	✓						
Implement your project									✓	✓	✓					
Monitor your project												✓	✓	✓	✓	
Publicise your success								✓								✓

KEY: ✓ Link or parallel

steps to identify potential difficulties and solutions, collect site and other data and information, and list possible ways to improve.

Legal and other requirements

In ISO 14001, procedures are required to identify legal and other obligations (such as codes of practice). The CP equivalents are similar to those for environmental aspects.

Objectives and targets

Organizations working to ISO 14001 must set some environmental objectives and targets. These must be written down and kept up to date. When setting objectives and targets, organizations must take account of their legal and other obligations, the significant environmental aspects they have identified, the views of interested parties and some practical business requirements. Equivalent steps in CP are setting project objectives and time frames, identifying and evaluating options and choosing projects.

Environmental management programme

ISO 14001 requires an organization to develop a management programme to achieve objectives and targets. It must set out who is responsible and how objectives and targets will be met. CP equivalents include steps to select teams, set project objectives and timeframes, identify potential difficulties and solutions, list areas with potential for improvement, draw information together, choose, prepare and implement projects.

Structure and responsibility

ISO 14001 requires organizations to define responsibilities for environmental management and provide resources needed to implement the EMS. CP equivalent steps are to provide resources (part of management commitment) and set up teams.

Training, awareness and competence

In ISO 14001, organizations must ensure that people carrying out activities that could affect the environment are competent. Training needs must be identified and a high level of basic environmental awareness training must be provided. CP equivalents could include obtaining management commitment (where training is needed), selecting teams and preparing projects.

Communication

Organisations working to ISO 14001 must have procedures for internal and external communication about their EMS and environmental aspects. CP equivalents could include team meetings and publicising successes.

EMS documentation and document control

ISO 14001 requires an organization to describe its EMS in formal documentation. This information must be kept up to date and procedures are required to control EMS documents. CP project information and management would appropriately be included in the EMS documentation and controlled by the document-control processes.

Operational control

Organizations working to ISO 14001 must identify operations and activities that are associated with significant environmental aspects of their activities, products and services. These operations and activities must be controlled so that their environmental impacts are managed. CP equivalents are steps to prepare and implement projects.

Emergency preparedness

In an ISO 14001 EMS, potential emergency situations must be identified. Formal procedures are required for responding to these and for preventing and mitigating environmental impacts. CP has a partial equivalent in the step to identify potential difficulties and solutions.

Monitoring and measurement

ISO 14001 requires organizations to have procedures to monitor operational processes, progress towards targets and objectives and compliance with the law. CP methodology also requires projects to be monitored.

Non-conformance, corrective and preventive action

ISO 14001 requires organizations to have procedures for dealing with problems in their environmental management. These must cover how action is taken to investigate problems and to make sure they are solved and do not recur. CP has partial equivalents in steps to identify potential difficulties and solutions and to monitor projects.

Records

In ISO 14001, formal procedures are required for record keeping. Records must be appropriate to the EMS and the organization and demonstrate that the requirements of the standard are being met. In CP, records would be generated through most steps of the methodology and would be an important part of monitoring projects.

Audits

Procedures are required by ISO 14001 for auditing an organization's EMS. All elements of the EMS must be regularly audited to ensure that planned arrangements and the requirements of the ISO standard are being met. The closest CP equivalent is monitoring projects, although formal, comprehensive auditing may not be part of the monitoring expected in the CP methodology.

Management reviews

ISO 14001 requires top management in an organization periodically to review the suitability and effectiveness of their EMS. In doing this review, the organization has to consider changing circumstances, the results of its own audits and, in particular, its commitment to continual improvement. The closest CP equivalent is monitoring projects. However, like auditing, formal comprehensive review may not be part of the monitoring anticipated in the CP methodology.

LINKS IN PRACTICE

On a practical level, New Zealand has some experience with linking CP with EMS. In 1994, 16 of the companies involved in the CP demonstration projects and published case studies were contacted to find out if such links were being made. These calls revealed some interesting information. Some were still working on their demonstration project, but none of the 16 had "formal", ongoing CP programmes. Ten were involved in other environmental management initiatives and projects which are linked to core business systems. Some were developing formal EMS, were involved in audits and projects to obtain new resource consents, or were developing programmes as part of business plans or total quality management initiatives.

Only half of the participants had heard of BS 7750, but four were using it as a model for their EMS. Thirteen of the participants had in place, or were developing, ISO 9000 quality management systems. Most of these had developed or were planning to develop links between the CP and environmental work and the ISO quality management system by including environmental requirements in procedures and work instructions. Two of the participants suggested that better links could be drawn with total quality management approaches.

Two years later (although a follow-up survey was not able to be completed) some of the companies have made progress in establishing formal EMS modelled on ISO 14001.

CONCLUSION

CP projects in New Zealand have demonstrated that substantial economic and environmental benefits can be obtained by businesses. In order to sustain these benefits and continue to create new benefits, New Zealand businesses have sought to link, or integrate, CP objectives into core business management systems. This integration is important, especially for small businesses, to maintain their efficiency, effectiveness and competitiveness.

EMS, such as those that are being developed to meet the requirements of ISO 14001 can provide significant support to promoting the objectives of CP. ISO 14001, requires a formal, documented EMS that is effectively implemented. It also requires a commitment and procedures to achieve continual improvement and pollution prevention. Even more important, ISO 14001 requires formal, comprehensive and ongoing monitoring, audit and management review. This should ensure that environmental management initiatives (including CP) remain on the senior management agenda and become part of ongoing business practice. ISO 14001, therefore, provides a framework for integrating CP into the core management of a business.

NOTES

1. Comments in this paper are based on the draft international standard, published in August 1995.
2. Q-Base is a quality management system code, based on ISO 9000, developed by Telarc NZ. Telarc provides independent certification to the code.

REFERENCES

Ministry for the Environment (NZ) (1993) *Cleaner Production at Work: Case Studies from New Zealand Industry*, July.
Ministry for the Environment (NZ) (undated) *Cleaner Production Guidelines*.

12
Cleaner production and environmental management systems in Australia

BRIAN O'NEILL

CLEANER PRODUCTION INITIATIVES IN AUSTRALIA

Environment leaders and environmental agencies in Australia recognize the potential for programmes (such as environmental management systems, cleaner production and eco-design) to act as mechanisms for building a culture of environmental best practice and environmental innovation throughout business and government.

In Australia, these initiatives in environmental protection became the focus of national programmes through the Australian Prime Minister's 1992 Statement on the Environment.

To assist Australian industry in improving its environmental performance, the Environment Protection Agency (EPA) administers a cleaner production programme as part of the implementation phase of that statement. The programme is geared to supporting industry initiatives to meet both increasingly demanding regulations and community concerns about the potential environmental impacts of industrial activities. More importantly, it is stimulating the development of an industry culture committed to achieving an environmental performance beyond minimum environmental regulatory requirements.

The programme is an Australian adaptation of work undertaken in the Netherlands. The objectives of the programme are both to increase awareness of the benefits of cleaner production and to encourage the uptake of cleaner production techniques and practices by Australian industries.

Environmental Management Systems and Cleaner Production, edited by R. Hillary.
© 1997 John Wiley & Sons Ltd.

AWARENESS RAISING INITIATIVES

In February 1993 the EPA released a Cleaner Production Information Kit to raise awareness of the advantages of cleaner production. More than 1200 kits have been distributed and have been well received by industry, industry associations and government.

Three series of cleaner production workshops have been conducted throughout Australia, one series in each of 1993, 1994 and 1995. The first two were conducted by industry representatives rather than government officials, while those held in 1995 provided a forum for Dr Don Huisingh to pass on details of his experience in cleaner production initiatives. Planning and implementation of the workshops required cooperation with state and local government, as well as business associations. As a result of the workshops, the EPA released in 1996 an Australian booklet, *Cleaner Production Case Studies*, along the lines of the popular United Nations booklet, *Cleaner Production Worldwide*. This booklet is free to companies, associations and international organizations. A database containing Australian cleaner production case studies can be found on the Internet at http://www.erin.gov.au/portfolio/epg/environet/nepd/case_studies.html as part of the Australian Environet initiative.

In line with its focus on facilitating the availability of information, especially to smaller and medium-sized enterprises, the EPA provided funding to a leading business association, the Australian Chamber of Manufactures, to ensure that its *Environmental Management Handbook* would be applicable in each Australian state and territory. This handbook provides a self-assessment procedure which small industries can use to assess voluntarily their environmental performance while also identifying the economic and environmental benefits of sound environmental management. The handbook has a very low price tag to encourage companies to use it.

DEMONSTRATION PROJECTS

National

The Australian government provided funding to a leading national trade union to produce a guide for its members on identifying and implementing environmental improvements in the workplace. The union has also run corresponding seminars for members and shop stewards and conducted pilot projects to highlight the importance of reducing environmental impacts. This will assist the workforce to look critically at their work areas from an environmental perspective.

The EPA has provided A$0.75 million for an EcoReDesign project in collaboration with industry and research institutions. Its goal is to promote to industry the potential of life cycle assessment and the advantages of incorporating environmental considerations at the design stages of products. The project commenced in June 1993. Funded by the Commonwealth Government, Eco-ReDesign is being coordinated by the

Royal Melbourne Institute of Technology. The first products from this project were launched during 1996.

Six Australian companies are already part of the project and are anticipated as providing strong and varied role models for the benefits of eco-design. An information kit will be produced and circulated widely.

The EPA is also administering a cleaner production demonstration programme based on the successful Dutch —Prisma— project. Selected companies are being assisted to undertake a review of their production processes and changes which result from these reviews will be used across Australia to demonstrate the potential benefits of cleaner production. The results will also be used to encourage industry (especially small to medium-sized enterprises) to adopt cleaner production practices.

Preliminary investigations indicate that some of the selected companies are likely to formalize their environmental management systems (EMS) as a means of assisting their cleaner production commitment.

The EPA has also developed a National Environment Industries Database (NEID). This details Australia's environment management capabilities in a range of sectors and has been available on the Internet since November 1995. An Australian National Cleaner Production Database which provides case studies and other cleaner production information to assist small to medium-sized enterprises (SMEs) is one element of the NEID.

Regional

The South Australian Cleaner Industries Demonstration Scheme is a joint project of the Australian and South Australian State government. It provides small grants and interest-free loans with a delayed payback period directly to companies for new technology. Funds are also made available for environmental audits to identify improvements in practices and technology which reduce costs and minimize pollution. This project complements the EPA's demonstration project while providing a focus on the specific needs of South Australia.

Other initiatives

A number of educational institutions now include environmental and cleaner production components in their mainstream courses.

ENVIRONMENTAL MANAGEMENT SYSTEMS IN AUSTRALIA

Through its AusIndustry programme, the Australian government provides financial assistance to SMEs to implement environmental management systems, with particular emphasis on cleaner production and waste management.

Companies worldwide are increasingly introducing environmental management systems into their corporate strategies. Standard-setting bodies, including the International Organization for Standardization (ISO), are giving high priority to the development of EMS standards. The EPA is following these developments with a particular interest in assessing their potential to assist in the implementation of cleaner production and improved levels of environmental protection.

A joint Standards Australia/Standards New Zealand Committee was formed in August 1992 to consider the need for an Australia/New Zealand EMS Standard. The committee (QR/11) has focused on providing input to the ISO process and on discussing Australian and New Zealand approaches at relevant ISO working group meetings.

International EMS standards can provide a model for use in incentive-based legislative approaches which reward companies for implementing systems to monitor, control and improve their environmental performance. They also have the potential to support national policies on sustainable development, while meeting free trade objectives. A number of Australian companies have already developed their own EMSs based on the British Standard BS 7750 and it is envisaged that some of these will convert these for certification in conformance with ISO 14001.

An international body established by agreement between the Australian and New Zealand governments operates as the peak accreditation organization of certification bodies in both countries. Known as JAS-ANZ (Joint Accreditation Scheme-Australia New Zealand), this Council has, as one of its objectives, the development of mutual recognition in overseas markets for Australian-certified products.

Between January and the end of June 1996 JAS-ANZ conducted a pilot programme to verify the appropriateness of its then draft environmental accreditation criteria and its accreditation for process for EMS certification bodies. This programme has been facilitated by the November 1995 two year adoption of the draft international standards ISO DIS 14001, 14004, 14010, 14011 and 14012 as interim Australian and New Zealand standards. In August 1996, the Commonwealth Government's Environment Minister presented certificates to four certification bodies, which had successfully come through the JAS-ANZ pilot programme, and ISO 14001 conformance certificates to 13 companies which had been clients of the successful certification bodies. At a ceremony in New Zealand certificates were presented to one certification body and two of its client companies. With accreditation and certification procedures firmly established in Australia and New Zealand, it is generally anticipated that there will be a steady increase in the number of accredited certification bodies and organisations with certified EMSs.

ENVIRONMENTAL MANAGEMENT SYSTEMS, ENVIRONMENT PROTECTION AND CLEANER PRODUCTION IN AUSTRALIA

It is reasonable to assume that organizations implementing an EMS are seeking a systematic framework for improving their environmental performance. Unfortunately, the adoption of this framework will not in itself achieve good environmental performance unless its objectives are derived from a soundly based environmental policy.

It is easy to lose sight of this fact, particularly given the conformance focus of ISO 14001 on core system elements. The importance of identifying sound environmental objectives, associated with review and performance indicators which attempt to measure both achieved levels of improved environment protection and the effectiveness of a system, cannot be overestimated. As the ISO 14000 series standards are designed to be applied in a wide range of organizations, environments, geographical locations and cultures, international EMS standards cannot, by definition, be too prescriptive about environmental outcomes.

If their potential is to be maximized and aspects of ecologically sustainable development expressed through cleaner production techniques are to be promoted, EMS in implementation need to include provision for continual improvement in environmental performance.

ENVIRONMENTAL MANAGEMENT SYSTEMS' EFFECTIVENESS AND OBJECTIVES

There is a concern, often expressed by community organizations, that some companies may only choose to set objectives which are limited to meeting their legislative requirements. A tripartite process for establishing an environmental policy which seeks input from the regulator, the community and the company is one way in which this concern can be minimized. While a proactive stance may initially entail increased cost or significant changes in long-established housekeeping procedures, it has been shown, in the long term, to lead to significant financial and environmental benefits and more acceptable environmental outcomes from a community perspective.

The level of environmental guidance given at the preparatory environmental review (PER) stage of each EMS is a significant factor in the quality of objectives for any resulting EMS. This raises important issues regarding the composition of auditing teams and cannot be neglected when considering the relationship of EMS to cleaner production and environmental protection. Australia is also interested in developments in the European Union voluntary Eco-management and Audit Scheme (EMAS).

As the concept of EMS develops, it may be necessary for particular industry sectors to provide guidance through such instruments as industry codes of best practice on environmental effects and performance levels in that sector.

CONCLUSION

From an Australian EPA perspective, an emphasis on performance outcomes is critical to ensuring that EMS standards genuinely promote environmental protection, rather than simply being certifications of commitment and compliance. Related to this is the importance of certifying bodies interpreting EMS standards in a performance sense as well as from a systems perspective.

Unless the system goes beyond mere record keeping or the implementation of process tasks, its potential to promote continued environmental improvement will be extremely limited. In this regard, the credibility of EMS standards hangs on their ability to deliver improved environmental protection outcomes beyond those needed for compliance.

There is also a concern that conforming to EMS standards may become so complex that it becomes inappropriate for consideration by SMEs. This issue will certainly be one which Australia will be watching with interest as these standards gain prominence in day-to-day business transactions.

Section III
Regulation or Self-regulation?

13
Introduction

RUTH HILLARY

This section of *Environmental Management Systems and Cleaner Production* considers the regulatory and self-regulatory mechanisms used to deliver cleaner production and improved environmental performance in enterprises. In six chapters, this section addresses the gradual shift in European Union (EU) environmental policy towards market-based instruments (Chapter 14); the next paradigm of environmental protection beyond compliance (Chapter 15); a regulator's view of a traditional pollution control system (Chapter 16); the elements of a formalized management system BS 7750 (Chapter 17); the value and credibility of accredited certification assessment (Chapter 18) and the efforts to harmonize national accreditation systems (Chapter 19). The central theme analysed throughout this section is the balance between regulatory and self-regulatory mechanisms and the credibility each system has in the marketplace to deliver sustained environmental performance improvements.

In Chapter 14, Ruth Hillary traces the development of European Union environmental policy as described in its action programmes, showing the heavy reliance on traditional normative legislation. She draws out the advantages and disadvantages of traditional regulation, asserting that the current shift in EU environmental policy is, in part, due to the failure of the past 20 years of legislative activity. Hillary proposes a model to describe the relationship between the regulator and the regulated which, she argues, incorrectly transfers responsibility for protecting the environment from the polluter to the enforcer. Furthermore, she claims that the shift towards market-based instruments is an attempt by EU policy makers to utilize an innovative mechanism to internalize the external costs to the environment of production. She considers two market-based tools: the eco-labelling regulation and the Eco-management and Audit Scheme (EMAS). Focusing largely on EMAS, she discusses its potential to deliver improvements in environmental performance, drawing on industrial examples to illustrate how existing company policies are falling woefully short of EMAS requirements. She analyses the economically viable application of best

---

Environmental Management Systems and Cleaner Production, edited by R. Hillary.
© 1997 John Wiley & Sons Ltd.

available technology (EVABAT) requirements of EMAS, suggesting the limitations of achieving EVABAT for small and medium-sized enterprises (SMEs). Hillary concludes that normative legislation will remain the driving force of EU policy and environmental improvements but that the challenge for both regulators and industry alike is to think strategically and holistically about how to achieve further environmental performance improvements.

David Rejeski, in chapter 15, continues the theme of the limitations of existing environmental policy raised by Hillary but in the US context. He argues that the next paradigm of environmental protection will be beyond compliance and based on voluntary risk-based environmental management. He considers the changing environmental agenda of a growing number of companies which have adopted a proactive, preventive approach to environmental management, calling it a quiet revolution. He argues that this revolution has been misunderstood or missed by US government officials for the past 25 years. Rejeski goes on to assert that cleaner production is dependent on the ability of organizations to learn and also that environmental clean-up and control are not only expensive, but also signal a basic failure in organizations and society. He cites virtual prototyping i.e. predictive modelling of complex systems, environmental management across the value chain and the evolution of customer-driven production as important changes in production. Rejeski argues that the command-and-control system becomes increasingly ineffective and potentially counter-productive as changes are realized in the production system. He proposes four changes to US environmental policy, concluding that these are necessary to unlock 25 years of regulation and adversarial behaviour so that government can move towards an innovation-friendly policy framework which facilitates the building of a more environmentally sustainable economy.

Allan Duncan, in Chapter 16, sets out the opposite position to Rejeski, bringing the debate back squarely into the forum of a traditional pollution-regulatory system's ability to require continual environmental performance improvements. He presents a UK regulator's view of voluntary environmental management systems, asserting that three uncertainties exist about such systems' role as a force for environmental improvement. The three uncertainties are SMEs' ability to adopt the systems approach; the capacity of such systems to deliver regional or national environmental targets; and the means by which such systems initiate, disseminate and make available technological development. Duncan then goes on to discusses the regulatory systems' strengths and weakness. He focuses, in particular, on integrated pollution control (IPC) and best available techniques not entailing excessive cost (BATNEEC) and their inbuilt mechanisms to deliver improvements in processes. Duncan concludes that the route to cleaner processes and a cleaner environment is through a combination of regulatory and environmental management systems tools set against a backdrop of partnership and shared objectives by all stakeholders.

Still within the UK context, Christopher Sheldon, in Chapter 17, provides an insight into the development of the first voluntary environmental management system (EMS) standard in the world: the British standard BS 7750. Sheldon describes the core purpose of BS 7750 as a tool for organizations seeking to improve their environmental performance. He also relates BS 7750 to environmental auditing, arguing that confusion surrounds auditing terminology. Sheldon considers the influence which BS 7750 has had on the EU developments of EMAS and the international developments of ISO 14001. He asserts that EMS standards such as BS 7750 help managers to pinpoint and manage their business's risks which, in turn, affect the efficient performance of their organization, therefore BS 7750 also helps safeguard their organization's access to and share of its market. But he concludes that a formalized EMS is only half the picture for an organization and that accredited certification by an independent third party is necessary for the credibility of an organization's EMS in the marketplace.

Jeff Dowson, in Chapter 18, amplifies Sheldon's theme concerning the credibility of a company's EMS being partly dependent on accredited certification by providing a detailed analysis of an assessor's role when certifying a EMS as conforming to either BS 7750 or ISO 14001. Dowson considers three stages in the assessment process—initial assessment, main assessment and surveillance visits—focusing, in particular, on the key issues related to the environmental effects/aspects evaluation and the determination of significance. He discusses the areas of concern often raised by organizations implementing formalized environmental management systems and seeking certification. Dowson argues that many organizations hold the misconception that quality auditors are able to transfer their skills to audit environmental management systems without further training. He concludes that environmental management systems certification is still in its infancy; consequently, the interpretation of the EMS standards is not fixed as it is with standards in the ISO 9000 series, but that accreditation criteria provide some guidance on how to build and integrate systems.

In the final chapter in this section, Roger Brockway takes the issues of accredited certification into the national and international arena. He argues that the function of accreditation is to ensure that independent certification can be trusted by the market and that national accreditation systems, through international harmonization, will ensure that certificates are understood and accepted globally. Brockway asserts that it is the job of accreditation to ensure that third-party certification is independent, competent and consistent. He contrasts the auditing requirements of quality management systems, which are focused on compliance with customer requirements, with those of the EMS ISO 14001, which is focused on performance. Brockway supports Dowson's ideas by asserting that a considerable amount of work is needed to adapt ISO 9000 auditors and certification bodies to undertake EMS auditing and certification. Although accreditation operates primarily at a national level, Brockway is concerned with the wide recognition of certificates and argues for the harmonization of criteria used by national accreditation systems as a way of facilitating the broad acceptance of certificates. He describes European and international efforts by the European Accreditation of Certification (EAC) and the International Accreditation Forum (IAF) respectively to bring together differing accreditation systems. Brockway concludes that certification is not a passive bureaucratic activity and that accreditation aims to add value by assisting companies to bring about environmental improvements which will satisfy their stakeholders.

Throughout the section a number of themes are developed by the authors, the main one being the determination of the optimal balance between regulation and self-regulation to facilitate the promotion of improved environmental performance in industry. The three central themes discussed in this section are:

- The advantages and disadvantages of traditional regulatory systems and the relative merits of other instruments, such as voluntary environmental management systems, to deliver environmental improvements.
- The use of environmental management systems to deliver environmental improvements and the mechanisms to promote their credibility in the marketplace by utilizing harmonized accredited certification.
- The relative effectiveness of regulation or voluntary environmental management systems to promote new technology.

A key feature of the section is its analysis of the implications of voluntary and mandatory systems for businesses and how these systems incorporate stakeholders' objectives and attempt to deliver a more environmentally sustainable economy. More detailed practical industrial experiences are presented in the following three sections to show how individual enterprises and industry sectors attempt to deliver improvements in environmental performance while addressing the views of their stakeholders against a background of very different legislative frameworks.

14
EU environmental policy, voluntary mechanisms and the Eco-managament and Audit Scheme

Ruth Hillary

INTRODUCTION

The Eco-management and Audit Scheme concept is not new. For many years, industrial and business associations and individual enterprises have developed internal environmental auditing systems to control their environmental impacts.[1] The idea for a pan-European eco-audit scheme was announced in the European Union's (EU) Fifth Action Programme on the Environment, "Towards Sustainability". Central to the scheme was the management tool of environmental auditing. The EU scheme offered potential advantages beyond existing approaches. It was designed to be a voluntary, market-based tool, independently assessed and therefore "credible". The scheme embodies the self-regulated approach supported by business organizations,[2] but few businesses are prepared to accept the degree of responsibility which goes with self-regulation.

The Eco-management and Audit Regulation (*Official Journal of the European Communities* 1993) was adopted on 29 June 1993 but only came into effect on 10 April 1995. The 21-month lead time allowed member states to develop the administrative structures and undertake trials of the scheme. It is these pilots, the experiences of the companies involved and the negotiations and attitudes surrounding the scheme which provide much of the data for this chapter.

Environmental Management Systems and Cleaner Production, edited by R. Hillary.
© 1997 John Wiley & Sons Ltd.

EU ENVIRONMENTAL POLICY

EU environmental policy is set out in its action programmes on the environment. Since the launch of the EU's first action programme on the environment in 1973, the Union has experienced growth not only in environmental awareness among the general public, governments and industry but also in the quantity and coverage of environmental legislation both at a national and Union level. The the EU has played an important role in this growth by defining environmental legislation. Over the past two decades, the EU has established four subsequent action programmes on the environment and more than 350 pieces of environmental legislation covering all spectrum of media: air, water and land, many products and processes, nature protection and environmental impact assessment (Commission of the European Communities 1993).

NORMATIVE LEGISLATION

The majority of EU environmental policy is based on the traditional normative or "command-and-control" type of legislation, which is typified by controls of discharges to atmosphere, water or land and environmental quality standards. "Command and control" succinctly describes the operation of such environmental legislation, because while the legislation sets the environmental standards for a certain process or operation—the "command" part—inspection is necessary by enforcement authorities to ensure compliance to the standards—the "control" part.

Normative legislation is necessary and has certain important properties in protecting the environment. Such traditional legislation has the important benefits of establishing guaranteed and harmonized levels of protection across the 15 member states. Disadvantages, however, do exist in this style of legislation. Frequently, only minimum standards can be set by normative legislation and these often quickly become out of date as new scientific evidence expands our understanding of environmental issues. Also, such legislation can not cover all environmental effects (which are numerous and increasingly complex as relationships between humans and their environment are further investigated and understood) but is fragmented, addressing separately the different media of water, air and land.

Considerable resources are required for the control part of "command-and-control" legislation. Member states must invest both financial and human resources to monitor, inspect and sometimes punish those organizations subjected to traditional environmental legislation. Since resources are not infinite and the financial cake needs to be divided among many competing interests, "control" cannot be comprehensive in any of the member states, although some are more effective than others. Similarly, member states have different degrees of expertise and resources to implement EU environmental legislation, resulting in a varied record of implementation of the different EU environmental laws (Department of Trade and Industry 1993).

Normative legislation has determined a distinctive relationship between the regulator and the regulated. An appropriate analogy would be the teacher/pupil discipline relationship. In this model, rules are set down by a governing body such as a national parliament and the EU, or the governing board in the case of a school, and the regulator, like the teacher, is in charge of making sure the rules are implemented and met by the regulated. The regulated enterprise, like the pupil, is expected to abide by the rules. What this relationship does to the balance of power between the two parties is to place the lion's share of responsibility for ensuring compliance on the shoulders of the regulator. The regulated enterprise, like the pupil, takes on the passive role, receiving external conditions on how it should act. In all schools there are good and bad pupils, similarly there is a spectrum of organizations and for the regulated there are powerful forces not to conform.

Non-conformance does occur. It can be due to a variety of reasons, including ignorance of the rules, more attractive alternatives arising from disobeying the rules or because the rules are impossible to meet—in the view of the regulated, the rules are not regulated effectively and/or the punishment is an insufficient deterrent. Although there is a clear responsibility for the regulated enterprise to comply with regulations, the option exists to disobey the rules, thus increasing the regulator's responsibility for ensuring compliance. Incentives to comply are negative, i.e. punishment and sanctions in the courts. The benefits of non-compliance are, among other things, saved expenditure and, if enforcement is minimal, the "good" organization is further penalized on its bottom line.

The regulator, as the enforcer of legislation, generally focuses on the regulated organization at the point at which it affects the environment, i.e. its emissions. Consequently, the regulator is the outsider. Furthermore, laws directed at outputs strengthen the view that the regulators have no role to play in the internal operations of the organization's attempts to achieve compliance. Increasingly, organizations are required to demonstrate their compliance by publishing to the regulators their methods to achieve compliance.

Regulators seek the manifestation of ability to achieve compliance to environmental regulations in technological terms, and companies respond with technological solutions. See, for example, the EU proposal for the Integrated Pollution Prevention Control Directive's Article 5, requiring enterprises to describe the technology and techniques used to reduce emissions. Member states' legislation also focuses on technological solutions.[3] Nevertheless, the relationship has directed efforts towards issue management with the solutions usually identified by technology.

TOWARDS SUSTAINABILITY

Union action programmes set out EU environmental policy and strategy but the Fifth Action Programme represented a shift in policy and strategy style from the four previous action programmes. This shift was in reaction to the continued and lamentable decline of the Union's environment, as detailed in the State of the Environment report (Commission of the European Communities 1992). Unlike previous programmes, which ran for a five-year period and leant heavily on the traditional policy instruments of normative legislation and financial aid, the Fifth Action Programme runs to the year 2000 and introduces three fundamental principles of sustainable development, preventive and precautionary action and shared responsibility, and a wider range of policy instruments.

Normative legislation will continue to play an important role in achieving the aims of the EU's Fifth Action Programme, but "Towards Sustainability" also proposes broadening the range of instruments to complement existing legislation. The new measures outlined are horizontal supporting instruments, financial support mechanisms and market based instruments. Five key sectors—industry, energy, transport, agriculture and tourism—are targeted in a coordinated and comprehensive way.

MARKET-BASED TOOLS

Market-based instruments offer probably the most innovative development in the Union's Fifth Action Programme. These are designed to internalize external environmental costs by alerting both producers and consumers to the need to use natural resources responsibly and so minimize and avoid pollution and waste. Essentially, market-based tools are about getting the prices right so that more environmentally friendly products and processes are rewarded in the marketplace. Market-based tools are expected to harness the creative energies of companies and direct them to improving the environmental performance of products and processes in a way which has remained untapped by the normative style of environmental legislation.

The Eco-labelling award for products and the Eco-management and Audit Scheme (EMAS) for processes are currently the only two EU market-based initiatives. Both are regulations and are designed to restore market forces in the environmental field by promoting competition on environmental grounds. As regulations they are directly applicable in all member states when adopted, but both are voluntary for company participation.

ECO-MANAGEMENT AND AUDIT SCHEME

During negotiations, opposition to the Eco-management and Audit Scheme regulation was considerable. Business associations opposed the regulation's concentration on defining internal management requirements, believing that these were not in the policy makers' domain. In contradiction to many associations' previous calls for self-regulation, a few made strenuous efforts to remove the draft regulation from the policy makers' agenda. More disappointingly for the future success of the scheme, many pan-European and national trade associations took no position at all. Many still do not have a position from which to advise their members.

Member states objected. Germany, in particular, objected to the generic nature of the scheme which, in its view, lacked specific environmental standards against which organizations would be judged. Germany stated that organizations located in a less stringent regulatory environment would achieve registration to the scheme and therefore be viewed as comparable to a company operating under a more stringent legislative regime, thus negating the latter company's achievements. Spain and Ireland maintained that the scheme was too bureaucratic for smaller enterprises, that it would be considered an unobtainable luxury for larger companies only, which they maintained were already well managed and therefore in less need of EMAS. Similar criticism has been levelled at the international quality standards in the ISO 9000 series.

The Eco-management and Audit Scheme regulation contains 21 articles and 5 annexes. EMAS is a voluntary, site-based scheme which requires participating companies to develop an environmental policy, undertake an initial environmental review at a site and develop an environmental programme in the light of the review's findings. It is necessary for the site to establish and implement an environmental management system (EMS), such as the British standard BS 7750 (British Standards Institution 1994) or the international standard ISO 14001 (British Standards Institution 1996), and to audit the system. Finally, the site must publish performance data in the form of an environmental statement and have this and its internal system verified by an accredited environmental verifier before succeeding in registering to EMAS (see Figure 14.1).

EMAS and national EMS standards

A provision for the recognition of national standards was included in the regulation under its Article 12 with the specific purpose of providing implementing organizations with an indirect route to achieving EMAS. Companies implementing and being certified to national, European or international standards will be deemed to have met those parts of the regulation as long as the standards used fulfil two conditions. First, the standards must be recognised by the Commission, and second, the standards must be certified by a body whose accreditation is recognized by the member state in which the site is located.

The existence of three national standards (the British Standard BS 7750, the Irish standard IS 310 and the twin Spanish standards UNE 77–801 and 77–802) which meet particular aspects of EMAS were agreed by a Commission decision on 2 February 1996. Therefore, in the case of UK enterprises, over 150 have been certified to BS 7750 and of the eight registered to EMAS by November 1995 all but one used their accredited certification to BS 7750 as a stepping stone to EMAS registration.

CEN MANDATE

To prevent the proliferation of national standards and to achieve one standard for Europe which could be used as a way towards EMAS registration, the Commission

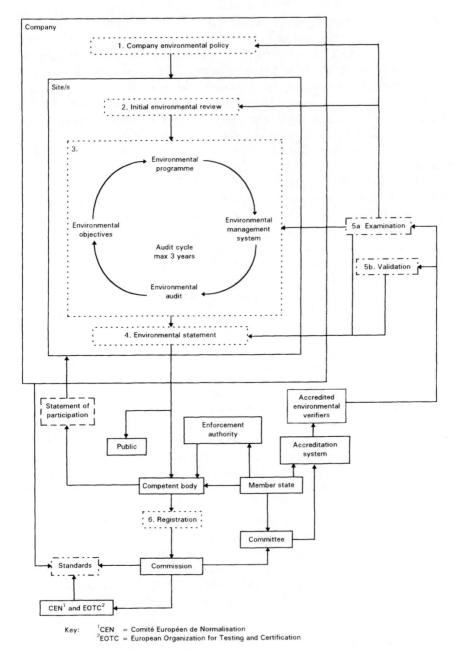

Figure 14.1 Implementation stages for the Eco-management and Audit Scheme (EMAS)

gave the European Standards Organization (CEN) a mandate to establish a standard for all aspects of EMAS except the environmental statement. CEN accepted this mandate in October 1994 and then had 18 months until April 1996 to prepare the standard. Aware of the activity in the International Organization for Standardization (ISO) Technical Committee TC 207 on environmental management, in particular its Sub-Committee 1 (SC1) on EMS which was in the process of developing ISO 14001, CEN decided not to duplicate these efforts but to participate fully in the ISO developments. CEN's intention was to work towards an EMS standard which could fit the purpose of its mandate.

Considerable efforts and speed have been expended on the development of ISO 14001. Voting on the draft standard ended on 10 February 1996. Comments were compiled by the UK secretariat of SC1 and a second two-month voting period was completed by the end of summer 1996. In September 1996 ISO 14001 became a full international standard. It is clear, however, that although strenuous efforts have been made to mould ISO 14001 into the shape of EMAS, the standard does not meet the CEN mandate of an environmental management systems standard compatible with EMAS (see Table 14.1). Consequently, three broad options are available to achieve compatibility: first, CEN could develop a bridging document to make ISO 14001 meet the mandate; second, ISO could produce an informative annex to the standard; and finally, the Commission could write a guidance note explaining how ISO 14001 needs to be expanded to meet aspects of EMAS.

The importance of achieving a good fit between ISO 14001, CEN's requirement for a European standard and EMAS arises because when a CEN standard is adopted all similar national standards must be removed; this is not the case when an ISO standard is adopted. Furthermore, CEN requires a standard that not only meets aspects of EMAS but also is not sharply different to an international standard which could raise the spectre of regionalization and trade barriers.

PILOT PROJECT COMPANY EXPERIENCE

The following description outlines the environmental policy element of EMAS, illustrating its potential to promote cleaner production and introduce a culture into an organization which broadens the organization's approach to its environmental issues. Practical experience is drawn from the preliminary findings of a jointly funded European Commission and UK Department of the Environment pilot project investigating the implementation of the scheme in small and medium-sized enterprises (SMEs) undertaken by the author.

The pilot involved 17 SMEs divided into three groups (see Table 14.2). The investigation started in March 1994 and ran for 12 months. The overall aim of the pilot project was to identify how SMEs may be assisted in the implementation process by identifying information and training needs and different supporting mechanisms.

Table 14.1 Comparision between the requirements of EMAS, ISO 14001 and BS 7750

Elements	EMAS (Articles and Annexes)	ISO 14001 (Clauses)	BS 7750 (Clauses)
Initial environmental review	Article 3b and Annex I.C	Annex A.3.1 (guidance only)	Annex A.1.2 (guidance only)
Environmental policy	Article 3b and Annexes I.A.1,2,3, I.B.1, I.C and I.D	4.2	4.2
Environmental management system	Article 3c and Annex I.B	4.1	4.1
Environmental programme	Articles 3c and e, Annexes I.A.5, I.B.1 and I.C	4.3.4	4.6
Environmental objectives and targets	Article 3e and Annexes I.A.4 and I.B.1	4.3.3	4.5
Environmental effects/aspects	Annex I.B.3	4.3.1	4.4
Environmental legislation	Annex I.B.3	4.3.2	4.4.3
Organization and personnel	Annex I.B.2	4.4.1, 4.4.2	4.3
Operational control	Annex I.B.4	4.4.6, 4.4.7, 4.5.1, 4.5.2	4.8
Manual and documentation	Annex I.B.5	4.4.4 and 4.4.5	4.7
Records	Annex I.B.5	4.5.3	4.9
EMS audits	Annex I.B.6	4.5.4	4.10
Environmental audit	Articles 3d and 4, Annexes I.C and II		
Management reviews	Annex I.B.1	4.6	4.11
Environmental statement	Articles 3f, 5 and Annex V	Not applicable	Not applicable

Table 14.3 SME profiles and environmental policies

No	Sector and NACE code	Employees Co.	Site	Environmental Policy	Review	ISO 9000	BS 7750
Group 1: sole companies							
1	Manufacture of furniture, NACE 36.14	34	34	Yes	No	In progress	No
2	Manufacture of insulation, NACE 26.14	154	144	Yes	Yes	Yes	No
3	Manufacture of metal joints, NACE 28.12	16	16	Yes	No	In progress	No
Group 2: Blackburn companies							
4	Manufacture of wallpaper, NACE 21.24	476	200	Yes	Yes, July 1992	Yes	In progress
5	Press felt, NACE 7.54		450	Yes	Yes, Oct 1993	Yes	In progress
6	Copper and aluminium powder, NACE 27.45	300	200	Yes	Yes, Oct 1992	Yes	In progress
7	Confectionery, NACE 15.84		150	Yes	Yes, Dec 1993	Yes	No
8	Geogrid, NACE 25.23	550	85	Yes	Yes, 1992	Yes	In progress
9	Rubber moulding, NACE 25.13	50	50	No	Yes, Dec 1993	Yes	No
Group 3: textile companies							
10	Dyeing and finishing, NACE 17.30	190	190	No	No	Yes	In progress
11	Dyeing, NACE 17.30	300	15	Yes	Yes	No	In progress
12	Dyeing and finishing, NACE 17.30	25	25	Yes	No	No	No
13	Dyeing and bleaching, NACE 17.30	86	86	No	Yes, March 1994	In progress	No
14	Dyeing and finishing, NACE 17.30	187	187	No	No	Yes	No
15	Wool spinning and weaving, NACE 17.12 and 17.22	150	150	No	No	No	No
16	Manufacture of wool products, NACE 17.73	279	94	No	No	No	No
17	Wool spinning and weaving, NACE 17.12 and 17.22	236	236	No	Yes, Autumn 1993	Yes	No

wrote an environmental policy. For most, it was the first time they had taken a strategic overview of the environmental aspects of their business. Working through a checklist of environmental issues, managers were directed to consider issues beyond those covered by their company's consents or licences.

Table 14.2 Pilot study groups

Group number	Group organization
Group 1	Three companies from various locations and sectors implementing the scheme on their own without external assistance
Group 2	Six companies located in Blackburn, north-west England, implementing the scheme with assistance from a new local Business and Environment Association
Group 3	Eight textile companies implementing the scheme with assistance from their traditional industry associations using sector application guides

Environmental policy

The central principle of EMAS is to achieve continuous environmental performance improvements of industrial activities.[4] Companies are required to make the twin policy commitments to comply with all relevant environmental legislation, both national and EU, and to achieve continuous environmental performance improvements. The purpose of the policy is to document the vision of where the company wants to be and secure top management commitment. A policy must be adopted and signed and dated by the chief executive officer or managing director. It must be publicly available and communicated and understood throughout the organization.

A number of surveys have already identified that only a small percentage of SMEs define their environmental strategy in an environmental policy.[5] Of the 17 companies involved in the EMAS pilot, 10 stated they had environmental policies (see Table 14.3). On closer examination, none fully conformed to the regulation's requirements as set out in Annex 1.C. Only five made a commitment to continual improvement and six to comply with legislation. Most companies originated their policies in response to an external enquiry, usually from a customer. One textile SME developed its environmental policy to display at a trade fair, whereas another SME's policy was in response to a new company chairperson who had a particular interest in the environment.

Environmental issue coverage in each policy was determined largely by the knowledge of the author of the policy and what he or she could glean from other companies' environmental polices. (This practice is not atypical of numerous companies.) One textile SME had simply taken another company's environmental policy and put its own name to it, even though the original company policy was in a completely unrelated sector and focused on waste management, which was a relatively minor consideration for the textile firm.

In Group 3, comprising of textile SMEs, none mentioned water management or savings, although one of their major resource usages is water. After individual discussions with each SME's management representative, each firm either redrafted or

Nevertheless, managers still had environmental "blind spots". Product planning, visual impact of the site and the environmental performance of contractors and suppliers failed to be considered in many policies (see Table 14.4). Water management and savings also remained low. Reasons for the blind spots varied. Product planning including design, packaging, transportation, use and disposal was neglected because products was perceived to have little environmental impact. Managers had not made the connection between the way they designed their products and the potential for minimizing environmental impacts downstream. Furthermore, product use and disposal were outside the site boundary and therefore perceived as not being a managerial responsibility unless there was a complaint about the quality of the product.

Table 14.4 Environmental issues covered in 11 pilot SMEs' policies

Topic	Present	Absent/weak
Water pollution	9	2
Soil pollution	10	1
Air pollution	11	
Reduction of impact on the environment	10	1
Energy management	8	3
Raw materials management and savings	9	2
Water management and savings	6	5
Solid and hazardous waste management	10	1
Working conditions (noise, odour, dust)	7	4
Visual impact	4	7
Product planning	5	6
Changes to process/selection of new processes	10	1
Contractors, subcontractors and suppliers	6	5
Prevention of major environmental incidents	7	4
Contingency procedures for major accidents	8	3
Staff environmental information and training	11	
External environmental information	9	2
Legislative compliance	10	1
Continual improvement	8	3
Target setting	11	
Policy review	8	3

In a similar way, contractors and suppliers were viewed as independent of the company and therefore difficult to influence. Three of the textile SMEs, however, had already been in contact with chemical suppliers to find alternative chemical products because of pressure from the regulator to eliminate chrome in trade effluent and formaldehyde in air emissions from stenters.

Although a number of sites were either located in beautiful areas or surrounded by housing, visual impact remained a low priority in their policies. Managers judged that they could do little about the appearance of their sites and/or that the local community had often developed around the site and therefore become accustomed to the site's appearance. Other companies were located in industrial estates and consequently considered that the whole area was expected to look relatively unattractive. One SME in Blackburn had made a concerted effort to improve its appearance, originally because its managing director was a member of a local environment initiative called Groundwork, but subsequently because the MD identified that improvements in the site's appearance increased employee morale and pride in the company and gave a better impression to customers and visitors.

Water management and savings continued to be omitted in a number of policies, even though discharges to water were considered very important where effluent was regulated by consents. Water usage was not a major priority because, first, many sites had their own bores and the cost of abstraction licences was minimal (a few hundred pounds sterling) and, consequently, water was considered as almost a free resource; and second, because reducing water consumption was believed to be either almost impossible or detrimental to product quality. Few companies had considered water consumption in conjunction with their effluent charges.

EVABAT

In the final stages of negotiations to adopt the regulation, a last-minute addition was included in its environmental policy, Article 3.a. The addition states that the policy commitment to environmental performance improvement should aim to reduce "environmental impacts to levels not exceeding those exceeding the *economically viable application of best available technology*" (EVABAT) (Author's emphasis). The insertion introduce a caveat to the regulation's principle of continuous environmental performance improvement. Nevertheless, it was introduced to enable the German delegation to lift their general reservation which had been blocking adoption of the scheme.

The German delegation were insistent that a reference to best available technology (BAT) should be included to enable environmental improvement to be benchmarked against technology, thus introducing environmental standards into the regulation. BAT was moderated by the UK delegation's insistence on including "economical viable application", limiting the extent to which a company would need to search for best available technology. In fact, the insertion is very badly worded and could be regarded as a way for companies to justify not setting ambitious targets. Furthermore, EVABAT inappropriately introduces into the policy the concept that all improvements are dependent on being assessed against technology.

Undoubtedly, technological solutions will be an important component of a company's environmental strategy as defined in its policy, but they form only one method of achieving environmental improvements. Others include focusing on raw materials, operator practices, training, redesigning products and working with customers. Emphasis on technology has the added disadvantage of reinforcing a predominant perception in industry that the environment means emission to media only and that these can only be tackled by the purchase of technology, usually "end-of-pipe".

Clean technology is referred to in Annex 1.D of the regulation in the context of 11 good management practices used to guide managers when they develop their company's environmental policy. These principles are also used to assess whether continuous improvement in environmental performance is being achieved. Managers are asked to consider the use of clean technology to reduce resource consumption and waste generation and emission. Clean technology is seen as one element to achieve environmental performance improvements, whereas EVABAT has the potential of constraining the strategic thinking required in EMAS.

CONCLUSION

EU environmental policy has broadened the range of instruments at the Union's disposal to develop European environmental strategy, moving away from the predominance on normative legislation towards market instruments. EMAS is a manifestation of this policy shift and offers the opportunity to harness market forces to stimulate companies to improve their environmental performance.

EMAS has not, however, had a spectacular start. The first five registrants appeared in the UK in August 1995. Over the following months, additional sites accrued slowly to the central Commission list, with the total registered sites topping 100 in January 1996. Although it is predicted that large numbers of sites will be registered to EMAS, its ultimate success, and that of other voluntary systems, will depend on market acceptance, which as most managers know is fickle and often unreliable.

Companies need to be encouraged to formulate environmental policies, as outlined in EMAS, to direct their efforts at all aspects of their environmental impacts as opposed to the emission-oriented focus that many currently adopt generally in response to normative legislation. While normative legislation will remain a driving force for improvements, the challenge for both regulators and industry is to think strategically and holistically about the environment. Both the regulated and the regulator will need to undergo a steep learning curve to enable them to understand fully the implications of committing to continuous environmental performance improvements.

NOTES

1. See, for example, organizational schemes such as the Chemical Industries Association's Responsible Care, the International Chamber of Commerce's (ICC) 16 Principles of Sustainable Development and the CEREAS re Valdez principle, and initiatives from individual organizations such as 3M's 3P programme and Allied Signal's environmental auditing programme which, in conjunction with a programme from Arthur D Little, forms the basis for the ICC environmental auditing handbook.
2. See, for example, the International Chamber of Commerce's comments on voluntary approaches to the environment and the Confederation of British Industry's papers on deregulation.
3. See "least polluting technology" in the Danish Environmental Protection Act, Ministry of the Environment 1992 and best available techniques not entailing excessive cost in the UK's environmental Protection Act 1990 which also outline integrated pollution control.
4. Industrial activities are listed under Section C (Mining and Quarrying) and Section D (Manufacturing) in the Community's statistical classification of economic activities in the EC, Council Regulation No 3037/90, OJ L293, 24.10.90: and electricity, gas, steam and hot water production and the recycling, treatment, destruction or disposal of solid or liquid waste.
5. See for example, Leicestershire Training Practice In Leicestershire Companies, UK which identified only 11% of SME had formal policies in 1993 and The Institute of Directors survey of its members which stated only 28% had policies.

REFERENCES

British Standards Institution (1994) *Environmental Management System BS 7750: 1994*, London, BSI.
British Standards Institution (1996) Implementation of ISO 14001:1996 Environmental Management Systems–Specification with guidance for use, London, BSI.
Commission of the European Communities (1992) The State of the Environment in the European Community, Com 92 Final, Vol. III, 27 March.
Commission of the European Communities (1993) "EC Legislation", *Environmental Research Newsletter*, No. 11, June.
Department of Trade and Industry (1993) *Review of the Implementation and Enforcement of EC Law in the UK*, an efficiency scrutiny report commissioned by the President of the Board of Trade, June, London, DTI.
Hillary, R. (1993) *The Eco-management and Audit Scheme: A Practical Guide*, Business and the Environment Practitioners Series, Cheltenham, Stanley Thornes Publishing.
Official Journal of the European Communities (1993) "Council Regulation (EEC) No 1836/93 of 29 June 1993 allowing voluntary participation by companies in the industrial sector in a Community Eco-management and audit system", L168, Vol. 36, 10 July.

15
Clean production and the post-command-and-control paradigm

DAVID REJESKI[1]

INTRODUCTION

It is one of the hallmarks of change that its outlines are easier to discern in retrospect. Those caught in the throes of a transition seldom see the contours of the new order. So it is with our existing environmental paradigm. Twenty five years after Earth Day we find a rising wave of discontent with status quo environmental policy, mixed with sentimental jeremiads about the passing of the old order and a hopeful searching for something beyond the regulatory horizon.[2]

This end point was predictable. The limitations of existing environmental policy were inexorably tied to the inability of laws to direct technological and organizational change in the face of increasing social complexity and the weakness of generic command-and-control and enforcement strategies to guarantee compliance (see Sparrow 1994 and Orts 1995). Threat and coercion can work, but absorb enormous human and financial resources, undercut innovation and often disguise true performance behind irrelevant efficiency measures.

The next paradigm of environmental protection will be different. Those who dabble in predictions have already named this new "state of affairs", calling it beyond compliance, voluntary compliance, risk-based environmental management, sustainable development, industrial ecology, green planning or the green path. However, it is premature to impose definitions on a loose set of emerging and ill-defined concepts. Instead, we should take the approach of those confronted with the end of other epochal events such as modernism, the industrial state, and the Cold War. We are

---

Environmental Management Systems and Cleaner Production, edited by R. Hillary.
© 1997 John Wiley & Sons Ltd.

entering the post-command-and-control paradigm. It lacks sharp edges or clear contours but it will be discernable in our emerging production paradigm—the way we choose to produce our goods and services in the twenty-first century. Clean production will be the purview of those innovators and leaders who shape the production system, create its underlying knowledge base and manage its ongoing transformation. Clean production will be largely the purview of industry.

In March 1970, *Forbes* magazine published an article with the prophetic title "Opportunity, thy name is pollution". It has taken US companies a while to catch on, but it is happening. Peter Coors, CEO of Coors Brewing Company, recently noted that, "fundamentally all pollution is lost profit". Speaking to the National Academy of Sciences, Paul Allaire, the CEO of Xerox, commented that, "There are good reasons to protect the earth . . . It's the safest and surest way to long-term profitability." 3M's "Pollution Prevention Pays" programme has cut overall emissions by more than a billion pounds since 1975 while saving $500 million, and Chevron has saved $10 million in waste disposal costs in the first three years of its "Save Money and Reduce Toxics" programme.

What is happening? A small but growing number of companies are taking strategic control of the environmental agenda. These firms have adopted a proactive, preventive approach to environmental management designed to move them ahead of growing regulatory burdens, environmental liabilities and negative public opinion. (This shift to proactive environmental management has been documented by a number of researchers, e.g. Hunt and Auster 1995; Fuelgraff and Reiche 1990; Piasecki 1995). Their numbers are small, certainly less than 5% of all firms, but it is in these companies, both large and small, that the outlines of the post-command-and-control paradigm are to be found.

This is a quiet revolution. It is likely to be missed or misunderstood by those in government who have spent the past 25 years wielding blunt instruments in the hope of getting industry to do their environmental bidding. Having spent so much time focusing on the cheaters and laggards in industry, public sector organizations are ill-prepared to understand the methods and motivations of the leaders. For over two decades, the learning rates of public sector environmental organizations have been linked to the slowest and most recalcitrant learners in industry, not to the leaders. It therefore comes as a surprise to many bureaucrats and policy makers when they stumble over a company which is defining its own environmental agenda without the prodding of government regulations or "incentives".

The greening of industrial leadership is no surprise, however, to those in business who are always seeking new ways to secure competitive advantage through product differentiation, technological innovation, first-mover advantage and better customer relations. For these companies, environmental excellence offers another way to step away from the pack and increase both the perceived and actual value of their products and services. It also allows these firms to move beyond the vagaries of uncertain or anachronistic environmental policies which are often threats or disincentives to strategic action in the marketplace.

It would be easy to dismiss many of these public pronouncements by industry as corporate rhetoric chasing green consumers. However, industry has developed, and is applying, a set of managerial approaches and technical tools needed to build the next generation of environmental improvements. Their motivation to apply these tools is economic self-interest and they will use them to raise the cost of competition in the marketplace. AT&T's philosophy captures this transformation. The company's focus on the environment has moved from being a matter of altruism, to competitive advantage, to competitive necessity.

In more and more companies, the drivers of environmental improvement are becoming internalized and the ability to solve environmental problems seen increasingly as a competitive asset with significant legal, financial and market implications (see Figure 15.1). In addition, the sphere of environmental management is expanding in both space and time, simultaneously increasing the range of opportunity for environmental improvement, productivity gains and profits. It may help to explore three shifts in corporate capability and thinking that will lay the groundwork for a new environmental paradigm. These changes will challenge the fundamental assumptions of environmental policy and, in so doing, provide a new set of very different challenges for the public sector.

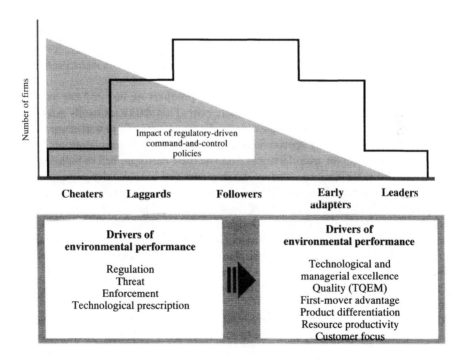

Figure 15.1 New drivers of environmental management

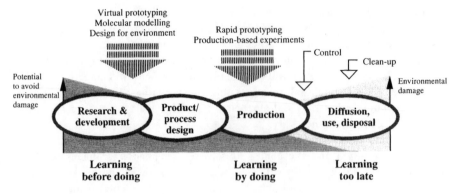

Figure 15.2 Shifting the focus of environmental learning

LEARNING BEFORE DOING

Ultimately, cleaner production will depend on the ability of organizations to speed up the rate of environmental learning and, to the extent which is possible, learn pre-emptively before damage is done. Environmental clean-up and control are not only expensive strategies, they signal a basic failure of our organizations and society to learn. For decades environmental engineers, scientists and policy makers have learned too late and environmental policy has been chasing production and its waste products (see Figure 15.2). This is about to end.

We are moving into a world where many of the products we buy and use will be designed, built and tested before they physically exist. This ability is broadly referred to as *virtual prototyping* and is made possible by the emergence of a new generation of computer-based tools. Using special supercomputers known as emulators or logic simulators, most computer manufacturers simulate their hardware before building it, allowing them to get a jump on software development and pinpoint design problems before large investment are made in production technology. It may come as a surprise to frequent flyers, but Boeing's new 777 aircraft was developed without the use of physical prototypes. Boeing engineers designed and assembled the plane in virtual space, allowing them to solve complex manufacturing and assembly problems early and do simulated walk-throughs of the airplane. Sikorsky flies a virtual prototype of its new military helicopter in what is known as a "synthetic environment", a computer-generated world which simulates weather, time of day and other mission-critical conditions.[3]

Why is this important? The predictive modelling of complex systems at different scales—from molecules to airplanes—radically alters the production paradigm and the opportunities for environmental improvement. In areas with robust scientific knowledge bases, such as electronics, aeronautics and chemistry, virtual prototyping is being used today to simulate new products, worker interfaces, operation and maintenance routines, and dramatically speed up the rate at which firms preemptively learn about the positive and/or negative effects of their decisions.

Virtual prototyping will allow environmental decision making to move upstream in time *before the manufacturing process*, before any waste or harmful pollutants flow into the environment. This capability allows us to begin to decouple environmental learning from actual physical production cycles and engage in what Gary Pisano at Harvard Business School calls learning before doing (see Pisano 1994 and Bohn 1987). The integration of environmental knowledge bases into computer-aided design tools is already occurring with systems such as Volvo's Environmental Priority System, AT&T's Green Index and IBM's Integrated Environmental Design System (described in Bell et al. 1995).

However, it will require time and resources to build integrated knowledge bases on materials and processes and capture this knowledge in virtual prototyping and other simulation systems. This is certainly one of the most important areas for cooperative industry–government research. In the meantime, significant environmental improvement can occur by speeding up our ability to learn about existing processes through techniques which increase experimental variety and capacity—essentially "learning by doing" faster. Dupont's nylon business, in Camden, South Carolina, has developed a "learning cell" to speed up the rate of experimentation around the production of nylon. Outfitted with over 100 sensors and a high sampling rate, learning from experiments reduced expensive interruptions in production and waste *by two-thirds* within a few months (after almost five years of marginal improvements).[4] The challenge is to turn such learning capabilities towards environmental issues and focus innovative organizational structures and methods on co-optimizing processes for better environmental performance.

ENVIRONMENTAL MANAGEMENT ACROSS THE VALUE CHAIN

As the temporal dimensions of environmental problem solving expand, the spatial dimensions are also shifting in ways which open up new possibilities for environmental improvement. Each firm is part of a long chain of value-creating activities, running from extraction of raw materials to fabrication, use, reuse, recycling and/or disposal. At one end of this chain are our natural resources, at the other our products and services. The territory in between is where the next generation of environmental improvements will be found. The explorers of this territory can create strategic insights and competitive advantage by understanding this value chain and the process, business and information linkages across it.

As corporations focus their managerial and accounting practices on the entire value chain, it expands the search for environmental opportunities beyond the boundaries of the individual firm to the larger enterprise (see Figure 15.3). This gives corporations a chance to move—up- or downstream—their focus on quality and continuous improvement, their internal cultures of technological and managerial excellence and their capacity for environmental problem solving. This process may become formalized through the mutual adaptation of environmental management systems and standards across the value chain (such as ISO 14001), or remain informal through inter-firm agreements, audits and environmental information sharing.

When Motorola was confronted with the task of eliminating chlorofluorocarbons (CFC) from its production processes, it turned to its supplier chain, often working with companies to transfer the required expertise and technologies upstream. Quad Graphics, a major printer of magazines such as *Time* and *Newsweek*, examined the environmental implications of its inputs from suppliers of paper and inks and is working to increase the eco-efficiency of the printing industry as

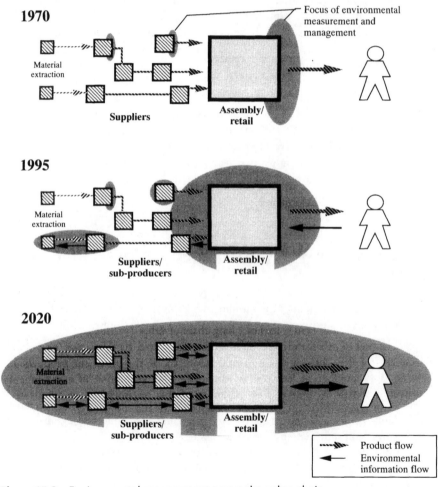

Figure 15.3 Environmental management across the value chain

industry as a whole. The 3M Corporation and Azko Nobel have developed an environmental guide to help one of the important purchasers of their products—the furniture industry—make better environmental decisions. Significant leveraging opportunities exist in the service sector which can use its enormous buying power to influence the environmental performance of its suppliers. Home Depot, a $12 billion hardware and home products chain, has been instrumental in stimulating companies making pressure-treated wood to examine the life cycle impacts of their products and production processes.

As environmental thinking becomes integrated into the value chain, corporations can move from building discrete solutions tackling firm-specific environmental problems to the integrated design of large technological and organizational systems capable of reaching much higher levels of environmental, energy and materials efficiency. As some researchers have pointed out (Womack and Jones 1994), moving from lean production to the lean enterprise allows breakthroughs to be "linked up and down the value chain . . . [so]the performance of the whole can be raised to a dramatically higher level". This allows companies to identify new opportunities to substitute intelligence, information and control systems for energy and materials and leverage the supply chain from raw material extraction to parts production and final assembly.

Value-chain thinking can potentially expand the boundaries of environmental learning and management, spreading environmentally sound practices across functional and firm-specific boundaries. However, this will not happen until environmental excellence becomes a management and financial issue instead of just an engineering or legal problem. Most companies have placed environmental management outside of the strategic core of the company where competitive advantage is built and most learning takes place.[5] As long as this is the case, the organizational learning rates for environmental improvement will lag far behind those affecting capital, labour and product value, and a wide variety of existing and emerging strategic management techniques will fail to be applied to environmental problem solving.

CUSTOMER-DRIVEN PRODUCTION

Downstream of the firm, another set of changes is taking place with far-reaching environmental consequences as firms continuously tighten linkages with customers. Innovative firms know that product improvements and profits depend heavily on speeding up learning cycles. This means assuring rapid feedback of customer preferences and values into the design and production process and making customers part of ever shorter product development cycles. The evolution of customer-driven production opens up new possibilities for customers' environmental values to move from the point of purchase backwards into the processes of production, design, development and even raw materials selection. This trend will be strengthened by the widespread adaption of flexible, computer integrated production systems and supply chain information management systems.[6] If customers demand more environmental information and value from producers, businesses can use this demand in the short term as a means of product differentiation and, over the long term, as a basis for building product and firm loyalty. The head of Patagonia, an environmentally conscious sportswear manufacturer, recently remarked that "we . . . rely on loyal customers to expand their definition of quality to include environmental responsibility."

How much are such strategies worth? Studies have shown that a 5% increase in customer loyalty can increase profits by an amazing 25–85% over the long term.[7] If corporate environmental stewardship can increase product rating and drive profitability and growth, companies will use it. A recent study showed that environmental brand enhancement allowed the makers of Arm & Hammer laundry detergent to realize an incremental $10 million in sales revenue for that product. Bryan Thomlison, former director of environmental management at Church and Dwight (makers of Arm & Hammer detergent) explains that we "[we] ratcheted up our environmental activity as a means of increasing our competitive advantage". A leading US manufacturer of wind turbines surveyed customers in the north-western states and discovered that 80% of the households polled would pay an additional $8 per month for electricity from environmentally clean sources, an increasing possibility in a deregulated electricity market. Providing green kilowatts to 1% of the residential electricity customers would generate $3–4 billion in potential sales. A recent Roper poll reports that almost 30% of the respondents have bought a product because of environmental advertising claims. Clearly, this is no longer a novelty market and more companies are viewing the environment as a profit driver rather than just a cost of doing business.

WHITHER PUBLIC POLICY?

It is important to realize that these changes in the production system, combined with a continual restructuring of core competencies in firms and expanding of the competitive domain, are not being motivated solely by environmental concerns, yet will have significant environmental implications over the long term. They will be complemented by more directed strategies such as full-cost accounting, life cycle assessment and design for environment. As the basic approach to the production of goods and services changes, and the temporal and spatial boundaries of environmental management expand, they will fundamentally reshape interactions between the public and private sectors on environmental issues. The command-and-control system will become increasingly ineffective, and potentially counter-productive, when applied to companies seeking sustained competitive advantage through process-embedded innovation, product differentiation, customer responsiveness and first-mover advantage. Government must adapt a more indirect role *vis-à-vis* corporate leaders as signaller, facilitator, and catalyst, with industry as the driver in the search for environmental innovation.[8]

Does this mean that government is *passé*? Hardly. Unfortunately, there will always be those companies that only respond to threats. It would be foolhardy, and probably impossible, to dismantle the existing regulatory system. As Michael Porter (1990) convincingly argues, environmental regulations have stimulated change and regulations, properly designed and applied, will continue to motivate some companies to improve environmental performance. Government will also play an important role in disseminating information on best environmental practices, conducting fundamental research, measuring and reporting on the state of the environment, ensuring scientific and technological literacy, and creating incentives to nudge the laggards to improve environmental performance. However, numerous interviews I have conducted with managers in many innovative U.S. firms have convinced me that the corporate environmental leaders need something additional from government, something which goes beyond "smarter" regulations and "pilot" projects. They require from the public sector a more general set of core competencies which characterizes a good business partner in a highly competitive environment.

First, they need more predictability. Unpredictable variations in policies, regulations, enforcement actions and standards suppress innovation, discourage investment, and cause market uncertainties. As firms and capital markets move faster, risks escalate and predictability in public policy becomes critical to overall strategic decision making in firms, especially small firms which often lack large financial cushions. One corporate executive I interviewed noted that "the most important thing the government can do for environmental innovation is to help facilitate the setting of clear, predictable long-term goals and policies". A recent review of the Environmental Protection Agency (EPA) by the National Academy of Public Administration emphasized the need for EPA to help set goals while providing flexibility to industry to meet these objectives (NAPA 1995). If goals are aggressive enough, they can motivate knowledge generation, stimulate learning and facilitate organizational change as well as pointing out technology and environmental performance gaps. In addition to goal setting, the public sector needs to ensure more predictability of innovative programmes over time so that government is perceived as a reliable partner and public–private sector programmes remain stable enough to produce results. As most entrepreneurs painfully know, money tends to run out long before the job is done. The government cannot afford to underresource policy innovation.

Second, environmental leadership will be facilitated by a more integrated policy system. At present in the USA, the EPA is governed by 12 separate statutes under the jurisdiction of dozens of separate congressional committees and sub-committees. This patchwork of requirements and mandates breeds a variety of ills including policy fragmentation, legal adhocracy and internecine battles between different policy-making and implementation bodies. Integration needs to take place at a variety of levels: geographically, between federal, state and local policies; thematically, between environmental, technology, economic and foreign policies; and temporally, across the life cycle of technological development from R&D to export. The existing mosaic of policies, programmes, and regulations undercuts efficiency, disaggregates markets and generally slows the ability to turn environmental innovation into environmental improvement, profits and competitive advantage for companies.

Third, public policy makers need to understand that time matters. One of the most dominant forces affecting global production systems is the shortening of product and process life cycles. As one corporate manager recently told me, "Time is the biggest single element in our business." One need only look around the nearest shopping mall to understand that value is increasingly sensitive to response time. This is what drives fast food, one-hour photo development and overnight mail. Companies such as 3M and Hewlett-Packard are generating 30–40% of their sales revenues from products introduced within the last two to three years. Seiko watches introduces 700 new products annually with its automated production line. Semiconductor manufacturers such as Intel are changing their chip production processes every 36 months. In six to eight months, the half-life of many government permit and grant applications, whole new product lines can be developed and launched. Without significant *and* intelligent increases in speed, the public sector risks becoming a drag on private sector innovation and competitiveness.[9]

Finally, the post-command-and-control paradigm must be built on a new social contract between the public and private sectors. Changing the environmental status quo will demand a level of openness and trust totally uncharacteristic of our regulatory command-and-control era. The challenge will be to create the required social capital—the norms, the collaborative networks and the trust—to support an open, participative system of environmental management with mutual benefits for both industry and the public. Certainly in the USA, the preconditions for this new social contract are not fortuitous, as recent research on civic engagement has pointed out.[10] The command-and-control paradigm asked little of its participants in terms of social skills and cooperation. In fact, it assumed asocial, defensive behaviour between largely anonymous parties and supported a legal and institutional framework which amplified blame, distrust and litigious behaviour. Slowly, over 25 years, we have backed ourselves into a Hobbesian nightmare where a tenuous agreement to avoid mutual harm has been built on fear and coercion rather than dialogue and trust.

What is at stake? Innovation has become internationalized and, much like capital, good ideas are free to move to friendlier shores. Many of the most environmentally innovative firms are multinational and can choose their location and markets. The goal of the public sector must be to build an innovation-friendly policy environment that attracts and holds the best ideas and idea generators. Nations or regions which can build a more predictable, integrated and time-sensitive policy framework achieve a first-mover advantage in the global competition for environmental innovation. This is not just an engineering or technological problem to be solved with more policy studies and R&D money. There are also social determinants of higher-level competitive advantage embedded in the way we manage organizations, learn and build social capital. In the end, these factors may determine who negotiates the tenuous path from an enforcement-driven, compliance-based approach to environmental protection to one which is performance driven and management based and, ultimately, who builds a more environmentally sustainable economy.

We are at a crossroads in attempts to define the post-command-and-control paradigm. This paradigm will define future markets, attract capital and drive environmental innovation far into the twenty-first century. Moving down the right road will not be easy. Momentum is a pervasive force in organizations and it tends to keep them moving in the same direction, often to dysfunctional extremes. After 25 years of regulation and adversarial relationships between the public and private sectors, we have "locked in" a set of mutually reinforcing competencies, institutions and behaviours which make change extremely difficult.[11]

Clearly, the post-command-and-control paradigm cannot be built by catering to tactical thinkers engaged in reactive behaviour. We need to look to where innovation is high and resistance to change low, to where responsible companies and creative public sector organizations have begun to apply a new set of tools and thinking to the next generation of environmental challenges. Here one finds reason for guarded optimism, but the path is not secure. As we reach out to grasp and shape our environmental future, we need to heed Ralph Waldo Emerson's warning of a century ago: "All things have two handles, beware of the wrong one."

NOTES

1. The ideas contained in this chapter do not necessarily represent the views of the Office of Science and Technology Policy or the United States government. The author can be contacted at: Dave_Rejeski@gnet.org.us.
2. A good overview of the present environmental policy debate in the United States can be found in *Environmental Science and Technology*, November 1995.
3. A virtual prototype has been defined as a "computer based simulation of a system or subsystem with a degree of functional realism that is comparable to that of a physical prototype". For a good overview of virtual prototyping in the aerospace and the defense sectors, see Garcia et al. 1994.
4. I am indebted to Jim Cook for providing me with information on Dupont's learning cell. This is an example of what has been broadly termed a learning laboratory, an organizational and technical system designed explicitly to enhance problem solving through purposeful experimentation and innovation. See Leonard-Barton 1995. The broad challenge is to design the production process itself to facilitate faster learning regarding key environmental and resource factors (materials, energy, etc.).
5. There is an inherent contradiction between command-and-control regulation and learning. By prescribing methods rather than end points, regulations suppress the primary tool used by organizations to learn—experimentation. See Rejeski 1995).
6. The value chain of many leading corporations has been "networked", allowing information on customer preferences to pass rapidly upstream. Benetton, the Italian clothes manufacturer, has linked its sales outlets directly to warehouses, flexible manufacturing facilities and the supply network, allowing the company to respond quickly to shifting market trends. Customers ordering Motorola pagers have their preferences captured on a bar code which drives a robotics line in Motorola's Florida manufacturing plant. Customers drive production variables enabling "lot-size-of-one" manufacturing. See Avishai and Taylor 1989.

7. The calculus of product loyalty is overwhelming. If you sell pizzas, a life-time loyal customer is worth $8000, if your product is Cadillacs, the value is over $300 000. Xerox calls its very satisfied and loyal customers apostles because of their tendency to repurchase Xerox equipment and convince others to do so. See Heskett et al. 1994.
8. A good discussion of the role of government in enhancing competitive advantage can be found in Porter 1990; and Porter and van der Linde 1995).
9. An EPA-funded study by the World Resources Institute came to a rather stark conclusion: "The great strides made by private institutions over the next thirty years will probably not be matched by public sector institutions. In the absence of fundamental change and renewal, the public sector could undercut progress in the private sector." See World Resources Institute 1993.
10. A discussion of the demise of social capital in the United States can be found in a number of recent papers by Harvard political scientist Robert Putnam. See Putnam 1995a and 1996.
11. Some researchers have pointed out the tendency to adapt and "lock in" suboptimal technologies even in the face of clearly superior alternatives. See David 1985. Less appreciated is that this problem applies to organizational behaviour as well, a phenomena know as "competence traps".

REFERENCES

Avischai, B. and Taylor, W. (1989) "Customers drive a technology-driven company: an interview with George Fisher", *Harvard Business Review*, Nov–Dec. pp. 107–114.
Bell, J., Bendz, D., Divakaruni, R., Hill, B., Mann, T. (1995) *Integrated Environmental Design System*, internal paper, Somers, NY, IBM Corp.
Bohn, R. (1987) "Learning by experimentation in manufacturing", available from the author, University of California, San Diego.
David, P. (1985) "Clio and the economics of QWERTY", *Economic History*, Vol. 75, No. 2. pp. 332–337.
Fuelgraff, G. and Reiche, J. (1990) "Proactive Umweltschautz" in Schenkel, W. and Storm, P. (eds) *Umwelt: Politik, Technik, Recht*, Berlin, Erich Schmidt Verlag.
Garcia, A.B., Gocke, R.P., Johnson, N.P. (1994) *Virtual Prototyping: Concept to Production*, Report DSMC 1992–93, Fort Belvoir, VA, Defense Systems Management College Press.
Heskett, J.L., Jones, T.O., Loveman, G.W., Sasser, E.W., Schlesinger, L.A. (1994) "Putting the service-profit chain to work", *Harvard Business Review*, Mar–Apr. pp. 164–174.
Hunt, C. and Auster, E. (1995) "Proactive environmental management: avoiding the toxic trap", in *The Best of MIT's Sloan Management Review*s in 1995, pp. 15–26.
Leonard-Barton, D. (1995) *Wellsprings of Knowledge: Building and Sustaining the Sources of Innovation*, Boston, MA, Harvard University Press.
NAPA (1995) *Setting Priorities, Getting Results: A New Direction for EPA*, Washinton, DC, National Academy of Public Administration.
Orts, E. (1995) "Reflexive environmental law", *Northwestern University Law Review*, Vol. 89, No. 4. pp. 1227–1339.
Piasecki, B. (1995) *Corporate Environmental Strategy: The Avalanche of Change Since Bhopal*, New York, John Wiley.
Pisaro, G. (1994) *Integrating Technical and Operating Knowledge: The Impact of Early Manufacturing Involvement on Process Development Performance*, Working paper 95-039, Boston, MA, Harvard Business School.
Porter, M. (1990) *The Competitive Advantage of Nations*, New York, Free Press.

Porter, M. and van der Linde, C. (1995) "Green *and* competitive: ending the stalemate", *Harvard Business Review*, Sept–Oct. pp. 120–134.

Putnam, R. (1995a) "Bowling alone: America's declining social captial", *Journal of Democracy*, January. pp. 65–78.

Putnam, R. (1996) "Americans losing trust in each other and institutions", *Washington Post*, 28 January, p. 1.

Rejeski, D. (1995) "The forgotten dimensions of sustainable development: organizational learning and change", *Corporate Environmental Strategy*, Summer. pp. 19–29.

Sparrow, M. (1994) *Imposing Duties: Government's Changing Approach to Compliance*, Westport, CT, Praeger.

Womack, J.P. and Jones, D.T. (1994) "From lean production to the lean enterprise", *Harvard Business Review*, Mar–Apr. pp. 93–103.

World Resources Institute (1993) *Focus Group Report: Institutions*, Washington, DC, WRI/EPA.

16

The role of regulatory systems in requiring cleaner processes and relationships with voluntary systems

ALLAN G. DUNCAN

INTRODUCTION

The purpose of this book is to assess environmental management systems to determine which promote cleaner production rather than the traditional pollution regulatory systems. Against this background it is worth noting the clearly implied reference to two types of system and recognizing their main features. There is a substantial and growing number of environmental management systems which are voluntary and usually apply at company or, more specifically, at site level; and there are those which are enforced by agents of government and apply nationally, and even internationally, in the context of the European Union for example.

The first is exemplified by the Eco-management and Audit Scheme (EMAS) introduced into member states of the European Union by way of a European Commission regulation but operating voluntarily at company level. There is also the system defined by the British Standards Institution (BSI) in BS 7750, which has analogues elsewhere. These systems are designed to be administered centrally and to include arrangements for formal assessment and certification of companies or sites which comply with the requirements of the system. At a somewhat less formal level there are systems typified by the British Chemical Industries Association (CIA)

Environmental Management Systems and Cleaner Production, edited by R. Hillary.
© 1997 Her Majesty's Inspectorate of Pollution.

"Responsible Care" programme; and there are systems for environmental reporting promoted by industrial groups such as the World Industry Council for the Environment (WICE) and the Public Environmental Reporting Initiative (PERI). All of these schemes require or propose the adoption of an environmental policy, an environmental management system and a statement about releases to the environment and about compliance with laws and regulations. The key theme which is common to all is an organizational commitment to continual improvement.

The second type of system is exemplified by the UK regulatory system for integrated pollution control (IPC) introduced by way of the Environmental Protection Act of 1990 (EPA 90). This statutory system is enforced in England and Wales by Her Majesty's Inspectorate of Pollution (HMIP) (on April 1st 1996 the Environment Agency superseded HMIP) and its main objectives are:

(a) to prevent or minimize the release of prescribed substances and to render harmless any such substances which are released;
(b) to develop an approach to pollution control that considers releases from industrial processes to all media in the context of the effect on the environment as a whole.

It has the following additional aims:

(c) to improve the efficiency and effectiveness of pollution controls on industry;
(d) to streamline and strengthen the regulatory system, clarify the roles and responsibilities of HMIP, other regulatory authorities and the firms they regulate;
(e) to contain the burden on industry, particularly by providing for a "one-stop shop" on pollution control for the potentially most seriously polluting processes;
(f) to provide the appropriate framework to encourage cleaner technologies and the minimization of waste;
(g) to maintain public confidence in the regulatory system through a clear and transparent system that is accessible and easy to understand and is clear and simple in operation;
(h) to provide a flexible framework that is capable of responding both to changing pollution abatement technology and to new knowledge on the effects of pollutants;
(i) to provide a means of fulfilling certain international obligations relating to environmental protection.

Although IPC is a relatively new concept, except in the context of regulation of radioactive waste management where it has been applied successfully in the UK for many years, it is enforced by way of a traditional regulatory system involving the issue of permits or authorizations for the operation of a process, followed by inspection of compliance with the conditions and limitations contained in the permit and implementation of sanctions for non-compliance. This system, too, has an inbuilt dynamic towards higher standards, albeit with qualification as discussed below.

This chapter deals primarily with the role of the latter type of system, but it may be worth noting a regulator's perception that the former type of system is, in principle, a considerable force for improvement. The voluntary commitment to continual improvement promises development and introduction of new techniques and progressive reduction of releases to the environment regardless of limits set by a regulatory system, and the publications of industrial groupings such as CIA, WICE and PERI, as well as those of certain financial institutions confirm that they are well apprised of the idea expressed recently by John Gummer, UK Secretary of State for the Environment, that: "Environmental excellence is not just something we should be seeking for its own sake, but also because it will increasingly be part of the price of access to the marketplace." There is also ample evidence that major industrial organizations, at least, appreciate that waste minimization and consequential reduction of releases are profitable, a concept expressed by slogans such as "Environmental Sense, Business Sense".

A good example of the influence of this type of system is the recent installation of activated sludge treatment of liquid effluent from an oil refinery operated by a major UK company.

Three doubts remain. The first is the question of how or whether small and medium-sized enterprises (SMEs) will respond. Julie Hill, director of the Green Alliance, has noted that they "may be less able to respond than larger companies. The overwhelming majority of UK companies fall into the SME category. Although it should not be assumed that they are therefore responsible for the majority of environmental problems, their aggregate effect on the environment is highly significant and their participation in environmental management initiatives of some kind is essential."

The second doubt concerns how or whether voluntary, company or site-related environmental management systems would necessarily deliver improvements required by regional or national environmental considerations, or by international obligations, without some kind of central administrative or regulatory system.

The third point is the simple but important question of just how technical development and innovation will be initiated and new, cleaner technology become available so that companies not in the business of process development can continue progressively to improve environmental performance. This turns out to be a question relevant also to the IPC regulatory system and is further discussed below.

THE REGULATORY SYSTEM: INTEGRATED POLLUTION CONTROL

The statutory basis for IPC is provided in Part I of the EPA 90 ("the Act"). It requires that no prescribed process may be operated without a prior authorization from HMIP (in England and Wales). Associated regulations prescribe the processes for control under the IPC system. These processes are those regarded as having the greatest potential for serious pollution of the three environmental media. (Other regulatory arrangements exist for control of processes with less potential for pollution. These are still single-medium related.) Further regulations define the procedures for applying to HMIP for authorization, the information required by

HMIP in an application, the bodies which HMIP must consult and the requirements for advertising the application and placing relevant information on public registers.

The system requires HMIP either to grant an authorization, subject to any conditions which the Act requires or empowers it to impose, or to refuse it. HMIP must refuse authorization unless it considers that the applicant will be able to carry on the process in compliance with the conditions of the authorization. In this regard, the Act places HMIP under a duty to ensure that certain objectives are met. These are as follows:

1. The best available techniques (both technology and operating practices) not entailing excessive cost (BATNEEC) are used to prevent or, if that is not practicable, to minimize the release of prescribed substances into the medium for which they are prescribed; and to render harmless both any prescribed substances which are released and any other substances which might cause harm.
2. When a process is likely to involve releases into more than one environmental medium (which will probably be the case in many processes prescribed for IPC), the best practical environmental option (BPEO) is achieved, i.e. the releases from the process are controlled through the use of BATNEEC to give the least overall affect on the environment as a whole.
3. Releases do not cause, or contribute to, the breach of any direction given by the Secretary of State to implement European Union or international obligations relating to environmental protection, or any statutory environmental quality standards (EQS) or objectives, or other statutory limits or requirements.

It is important to recognize the significance of IPC in regard to processes. An authorization under this system is for the *operation of a process*; it is not simply a consent to discharge substances into the environment. It is also relevant, in the context of improvement, to note that the Act requires authorizations to be reviewed not less frequently than every four years. The fundamental importance of the above objectives is reflected in the regulations relating to the information to be supplied by an applicant for an authorization. These require specific information about the matters on which the applicant relies to establish that these objectives will be achieved in operation of the process.

In practice, this means that, for a new process at least, the applicant company must show how, from the process design stage, it has selected and developed its proposed combination of primary process and pollution abatement plant having regard to the above objectives for its specific case. And HMIP must be satisfied with the arguments.

In passing, it is worth noting the importance of objective 3 above and recalling the question of how this might be met by voluntary, company or site-based environmental management systems. It is also worth considering the extent to which the matters of environmental quality standards or international obligations are themselves a force for development of cleaner processes. But in regard to the specific matter of promoting cleaner production it is necessary to focus on objective 1, the requirement to use BATNEEC.

BATNEEC

The document *Integrated Pollution Control: A Practical Guide*, published by the Department of Environment in 1996, gives the following definitions of "BAT" and "NEEC":

> *"BAT"*—It is helpful to consider the words "best available techniques" separately and together.
> "Best" must be taken to mean most effective in preventing, minimising or rendering harmless polluting releases. There may be more than one set of techniques that achieves comparable effectiveness—that is, there may be more than one set of "best" techniques.
> "Available" should be taken to mean procurable by the operator of the process in question. It does not imply that the technique has to be in general use, but it does require general accessibility. It includes a technique which has been developed (or proven) at pilot scale, provided this allows its implementation in the relevant industrial context with the necessary business confidence. It does not imply that sources outside the UK are "unavailable". Nor does it imply a competitive supply market. If there is a monopoly supplier the technique counts as being available provided that the operator can procure it.
> "Techniques" is defined in section 7(10) of the Act. The term embraces both the plant in which the process is carried on and how the process is operated. It should be taken to mean the components of which it is made up and the manner in which they are connected together to make the whole. It also includes matters such as numbers and qualifications of staff, working methods, training and supervision and also the design, construction, lay-out and maintenance of buildings, and will affect the concept and design of the process.
>
> *"NEEC"*—"Not entailing excessive cost" (NEEC) needs to be taken in two contexts, depending on whether it is applied to new processes or existing processes. Nevertheless, in all cases BAT can properly be modified by economic considerations where the costs of applying best available techniques would be excessive in relation to the nature of the industry and to the environmental protection to be achieved.
>
> [© Crown Copyright. Is reproduced with the permission of the Controller of Her Majesty's Stationery Office.]

For the purpose of assisting inspectors to determine what is BATNEEC in a specific case, HMIP publishes a series of *Chief Inspector's Guidance Notes* (CIGNs). These cover specific prescribed processes or classes of process and currently describe what is achievable with best practice in the relevant industrial context in regard to preventing, minimizing or rendering harmless polluting releases. In revision of these CIGNs it is planned to include information about the economics of classes of process and about what might be construed as the best practical environmental option (BPEO) in typical cases. The information is based on the results of extensive review of available techniques worldwide and is subject to wide consultation with interested parties within the UK.

Furthermore, the Chief Inspector of HMIP is under a duty by way of the Act "to follow developments in technology and techniques for preventing or reducing pollution of the environment due to releases of substances from prescribed processes". So CIGNs will be revised every four years, in line with the timescale for reviewing authorizations.

The judgement on BATNEEC by an inspector must have regard to the circumstances of a specific case, but it is generally expected that the standards described in CIGNs will be BATNEEC for **new processes**.

In relation to **existing processes**, HMIP must establish time scales over which they will be upgraded to new standards, or as near new standards as possible, or ultimately closed down and a schedule of improvements and time scales is generally included in an authorization. For this purpose too CIGNs will provide the guidance on standards.

In considering the role of BATNEEC in relation to the force for cleaner production, it is also important to understand the full significance of the term "techniques". Reference to the definition above shows that, as well as technology, it includes matters such as numbers and qualifications of staff, training, supervision etc. Experience shows that these factors have a substantial influence on, for example, the frequency and consequences of untoward incidents or of abnormal plant operation, both of which may have a substantial effect on the cleanliness of production.

A FORCE FOR CLEANER PRODUCTION?

In summary, we see that the IPC regulatory system offers the following:

- Operation of process requires prior authorization subject to conditions and limitations, including programme for improvement of existing processes.
- Sanctions for non-compliance with authorization.
- Conditions and limitations of authorization must deliver specific objectives, including use of BATNEEC, BPEO and compliance with EQS and international obligations.
- Current information available to inspectors, by way of CIGNs, for use in judging BATNEEC.
- Regular updating of CIGNs.
- "Techniques" in "BATNEEC" refers to highly relevant matters in addition to process technology.
- Statutory requirement to review authorizations not less frequently than every four years.

Against this background, it is contended that the IPC regulatory system is a very powerful force for improvement of processes and, indeed, of the environment as a whole.

It is still rather early to cite examples of major improvements actually delivered as a consequence of the system, but prospective improvements are shown in programmes included in authorizations placed on public registers. Nevertheless, a number of encouraging improvements have been made on smaller, more flexible processes as a result of the IPC system. At this stage of development of IPC they are necessarily associated with abatement technology rather than the primary, waste-producing process. They include examples such as reduction of cadmium releases from plating operations and of particulate releases from smokeless fuel production. As regards primary processes there are also encouraging indications in presentations made to HMIP on processes for glass production with lower generation of NO_x and for power generation by way of integrated fuel gasification, combined cycle gas turbine techniques giving improved systems for sulphur removal at source.

The system, however, is not without some weaknesses. In regard to the objective of "promoting cleaner production", attention must be drawn to the definition of "available" in BATNEEC. It requires that "techniques" should already be procurable by an operator even if only at pilot scale, provided that this allows its implementation with the necessary business confidence, i.e. at a fairly advanced stage of development and proving. This leaves open the question of how innovative techniques are initiated and developed to the extent that they are "available" for the purposes of BATNEEC. This same question was raised in regard to the voluntary, site-related environmental management systems, and is discussed further below.

There are other points of regulatory detail which may be worthy of discussion in the context of "promoting cleaner production". One example concerns the interpretation of BATNEEC and BPEO when a process may use a "waste" as a feedstock. The process may then be less clean than if it was using a cleaner feedstock, but the overall effect on the environment may be beneficial in the BPEO context. This is a question of where the boundary is drawn for consideration of the overall environmental effects. Presumably, in the site-related, voluntary system the environmentally sensitive operator would choose the "cleaner feedstock process", subject to economic practicalities. It might require need a system with a wider horizon, such as the regulatory system, to secure a process which operates in the interests of the wider environment.

INNOVATION AND TECHNOLOGY FORCING

It is contended that the IPC system is a powerful force for cleaner production, but it is recognized that its requirements cannot exceed what is "available". It does not have the power apparently available to the US regulatory system, for example, to set standards for the future which are not achievable with contemporary technology and therefore require innovation. The same limitation applies to the voluntary systems when a company committed to continuous improvement does not have the facilities for innovation or process development and so cannot improve beyond the best available. The promotion of cleaner production therefore requires attention to how innovation is initiated and processes developed to the extent of being "available". There would seem to be various factors or driving forces in this regard, including the following:

(a) *Encouragement by government:* For example, the Department of Environment and Department of Trade and Industry in the UK have recently launched an Environmental Technology Best Practice Programme. The R&D element of this programme provides financial assistance at a rate not exceeding 49% of eligible costs to stimulate the development of innovative environmental measures. The projects must be collaborative, with at least two UK organizations contributing to, or investing in, the work. (Funding of small, innovative projects by regulatory organizations, such as HMIP, may be regarded as a sub-set of this.)

(b) *Entrepreneurial activity by industry:* R&D is carried out by plant and process manufacturers who perceive financial advantage in developing novel processes for sale to customers committed to a programme of continual improvement. This activity may be further encouraged by the prospect of potential customers becoming exposed to economic incentives in regard to environmental performance or to pressures from tightened environmental quality standards or international obligations (e.g. under OSPAR or North Sea Conference decisions or recommendations). This activity may also be undertaken by operators who, themselves, may become subject to such pressures.

(c) *Studies by universities or research establishments:* This is a traditional source of innovation but it depends generally on funding from government or interested organizations including industry and non-governmental organizations (NGOs).

(d) *Influence by financial institutions:* Increasingly, the major banks are influencing environmental performance by way of banking practices which have regard to environmental policies. For example, the UNEP Advisory Committee on Banking and the Environment, which comprises a substantial number of banks worldwide, has produced a policy which commits them, among other things, to support and develop suitable banking products and services designed to promote environmental protection, where there is a sound business rationale. (This would seem to complement factor (b)).

CONCLUSION

It seems clear from the above that, where there is a common objective of continual environmental improvement, the most effective route to cleaner processes and a cleaner environment is by way of a combination of voluntary environmental management systems and regulatory systems. Their complementary features would seem to cater well for the variety of circumstances presented by large and small enterprises, by companies more and less committed to continual improvement and by the need to see the environment in a wide regional, national or even international context. It is also clear that this is best achieved through partnership in this objective by industry, including financial institutions, government, its environmental regulators and NGOs.

NOTE

It is acknowledged that the basis of this chapter has already been used in materials published by Her Majesty's Inspectorate of Pollution ©, and is reproduced with permission.

17
BS 7750 and certification—the UK experience

CHRISTOPHER SHELDON

INTRODUCTION

The British Standards Institution (BSI) has a long history of providing technical standards and related services to industry, and is perhaps best known for pioneering BS 5750, the national standard for quality management systems and the template for the ISO 9000 series of international standards. A Royal Charter body, independent of both UK industry and government, BSI has been responding to the needs of industry and consumer alike since the beginning of the twentieth century. Much of this is well known, but what may prove more surprising is that BSI has been in the forefront of environmental standards for over 40 years.

Combine this substantial track record with BSI's continued worldwide success in the field of quality management training and certification, its vital role in the development of the single European market and its role as a technical forum where issues can be debated and consensus established, and there can be very few who would dispute the Institution's current important role in global environmental developments.

As part of an environmental initiative designed to aid organizations in coping with a relatively new aspect of industrial life where there are few established work practices and processes, BSI published *BS 7750, Environmental Management Systems*. This document provides industry with a generic model that will help individual organizations to establish, develop and maintain their own purpose-built environmental management system (see Figure 17.1 and Box 17.1). The standard is not only a vital tool for those seeking to improve their environmental performance, it also provides an

---

Environmental Management Systems and Cleaner Production, edited by R. Hillary.
© 1997 John Wiley & Sons Ltd.

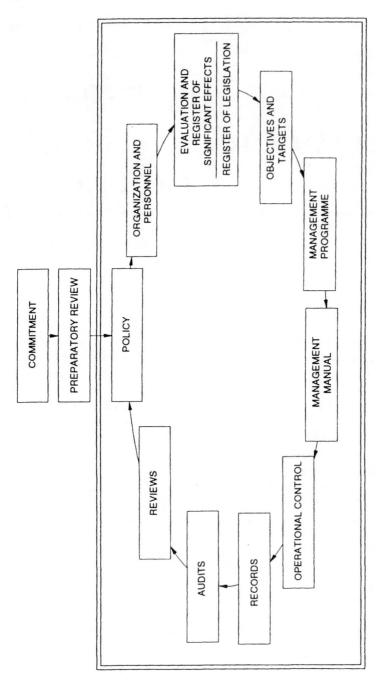

Figure 17.1 BS 7750 environmental management systems. Source: BSI 1994

Box 17.1 How the standard works

These are the basic building blocks specified in the standard for organizations which wish to set up their own environmental management systems.

Initial review: Sometimes inaccurately referred to as an environmental audit, the review should provide a detailed "snapshot" of the organization's environmental impacts and performance.

Policy: Drawn up in detail by the management team, the standard's only stipulated performance requirement is a "commitment to continual improvement".

Organization and personnel: The standard requires clearly defined and stated responsibilities, authorities and resources. It also covers policies on training and communication.

Register of regulations: The standard requires up-to-date records of all legislation that currently pertains to environmental aspects of the organization's activities.

Evaluation and register of effects: The organization needs to establish procedures to identify and evaluate all its environmental effects, direct and indirect, and compile a register of those considered "significant".

Objectives and targets: These need to be identified and communicated throughout the organization

Management programme: This should be drawn up to support the targets and should allocate responsibilities and provide details as to how targets should be achieved.

Management manual: This should collate the policy, objectives, targets and programme, and provide a key to the documentation providing the system.

Operational control: The organization should ensure that the control, verification, measurement and testing required to support the programme are carried out adequately.

Records: These should be detailed enough to show how the management system is working, and to record progress towards the objectives and targets.

Audits: These are to confirm that the policy is being carried out and that the programme is being adhered to.

Reviews: The organization will need to carry out a major review at periodic intervals to ensure that its management system conforms to the standard, and that may include changing the policy if objectives have been reached.

excellent framework for cost–benefit analysis to be carried out in an area where increased efficiency means greater profits.

WHEN IS AN AUDIT NOT AN AUDIT?

Without some background information, it may not be easy to see how such a standard could help organizations in this way. As the concept of environmental auditing has changed and broadened, an element of confusion has crept into the use of the term "environmental audit", although the International Chamber of Commerce (ICC) has defined an environmental audit as:

> A management tool comprising a systematic, documented, periodic and objective evaluation of how well environmental organization, management and equipment are performing with the aim of helping to safeguard the environment by (i) facilitating management control of environmental practices; (ii) assessing compliance with company policies, which would include meeting regulatory requirements.

Confusion is sometimes caused by the fact that the term implies a regular examination and verification of processes that are already established, whereas many companies are carrying out the examination for the first time, some with little knowledge of the systems and environmental effects they are supposed to be "auditing". Such companies are in reality carrying out a positional "review" rather than an audit.

The result of such a review may well emphasize that the company's processes lack structure and coherence in some, if not all, areas. One way to achieve greater proactive management of these processes is a systematic and integrated managerial approach, specified in BS 7750. Once such a management system is in place, it follows that any subsequent audits will have the maximum beneficial effect.

As previously mentioned, BSI has had extensive experience of standardizing management systems, having produced BS 5750 *Quality Systems* in 1979 and played a major role in its updating and eventual adoption as an International Standard (the ISO 9000 series in 1987). Now adopted by the European Committee for Standardization (CEN) as the EN 29 000 series, quality management system standards have obvious parallels with the requirements for environmental management. So much so that the UK again lead the way by chairing the technical committee writing an international version of an environmental management systems standard, known as ISO 14001, and published in September 1996.

THE EUROPEAN CONTEXT

In terms of other standards on environmental management systems, certification and implementation of BS 7750 have influenced not only international work but other European moves. CEN, the European standards body, has a long standing agreement that, in the absence of a European standard in an area, new work should be undertaken internationally, unless there is an overriding need for a European version. CEN has now been mandated by the European Commission to produce an environmental management system standard in support of the European Union's Eco-management and Audit Scheme (EMAS) regulation. Owing to the speed of development of ISO 14001, CEN has appointed a working group to develop a "bridging document" to allow companies which wish to use ISO 14001 in pursuit of EMAS, avoiding having to create a separate European standard. Until then, BS 7750 is one of three national standards recognized by the European Commission as meeting the management system requirements of EMAS.

Although there are strong links with the ISO 9000 series of standards, companies which are not currently using ISO 9000 are still able to use the environmental management system standard, as care has been taken to ensure that it is a "stand alone" document. However, there are some important additional requirements in the new document that mean it could not be considered simply as an extra part of ISO 9000.

Until recently, many organizations have focused on developments in European and national legislation; again, the standard can not only help in this area but has already influenced the enactment of such legislation. For instance, BSI has had representation on a special working group that is helping the European Commission with its plans to implement EMAS, which was launched in the UK on 10 April 1995. BS 7750 is not only compatible with the "environmental management system" requirements of the regulation, but is so influential that substantial pieces of the standard's text are quoted in the important advisory annexes, where guidance to those wishing to comply with the scheme is given.

At whatever level, compatibility of BS 7750 with developing standards and legislation was considered paramount. So mindful of this aspect was the technical committee that wrote the standard, that a review of BS 7750 was planned within 12 months of the original publication. In addition, a special 12-month pilot implementation programme in the UK assessed the effects of the standard in use and ran through until April 1993. Involving over 450 organizations spread across 38 different industry sectors, the information and experience gained from the programme was fed back into the review of the standard and resulted in a revised document being published in January 1994.

As with previous standards on management systems, BS 7750 is applicable to all types of organizations, whether manufacturing or service providing, no matter what their size in terms of workforce or turnover. Costs to the organizations which undertake implementation of the standard will obviously be dependent on a series of factors such as size, relative complexity of operations, the existence of any other standardized management systems, information and manpower resources, as well as any other specific environmental initiatives undertaken before BS 7750 implementation. The significant benefit is that not only is the cost of implementation within the control of the company, but also the result is greater cost control of an area that is increasingly threatening the balance sheets of all organizations through spiralling energy and waste-disposal costs.

One of the most important aspects of the standard is that it is designed as a management tool not a regulatory device, enabling management teams to devise their own policy and then provide the necessary support and information systems that are required. Assurance based on compliance with the standard is thus centred on the ability of the management to meet its own stated objectives, and on the actual level of performance attained in the achievement of those targets, which is addressed by the assessment of the organization's internal auditing process.

CERTIFICATION: THE WORKING PROOF

In effect, the development of BS 7750 is only half the story. The missing half is the development of credible accredited certification to the requirements of the standard by an independent third party. Following nearly a year's intensive work with a wide range of industry, BSI Quality Assurance now offers just such a service. Again, this was a world-leading development as, at the time of its launch in March 1995, no other country anywhere in the world had developed such a framework to help companies increase their efficiency and demonstrate their commitment to their customers.

Interest has already been shown around the world, notably from the USA, the Pacific Rim and South America. Obviously those companies with ISO 9000 systems already have a significant lead when it comes to implementation.

In order to aid understanding of the broad principles of the standard, a special breakdown of the basic requirements for certification has been included (see Box 17.1). From this it is relatively easy to understand how an established environmental auditing system, and many other internal industrial initiatives, fit in under the umbrella of the management system.

The challenge for many companies is that a new aspect of business life, the environment, is being entered on the balance sheet for the first time. Whether or not it considers it appropriate to enter the European EMAS scheme, using the new standard will certainly give a company (and its other clients and audiences) confidence in the knowledge that it is tackling environmental concerns in a systematic and integrated way. Not only that, but the company itself will reap the rewards of greater cost control and business process efficiency. Environmental management is no longer a "nice to have" but a significant weapon in the business manager's armoury in the fight to achieve maximum result for every penny spent.

Early action by far-sighted organizations has ensured that they are well placed to meet the requirements of the international standard ISO 14001, the impact of which on global trading is still being calculated. BS 7750 is the best tool currently available for all managers who want to be sure their organization is achieving the best it can environmentally. BSI QA's experience in certification and assessment of such systems ensures that the latest in best practice is disseminated to all users of the standards who seek to ensure that their customers know of their current efforts and future aspirations in the field of environmental performance.

Certain companies did not wait for the publication of the final international standard, but pursued accredited certification to the draft version. Companies such as these have not surrendered the bottom line as their top priority, but neither do they agree that the environment is a side issue. Instead they have begun the difficult task of integrating environmental considerations into their mainstream managerial activities, and future practitioners of environmental management have much to learn from these pioneers.

CHANGING MANAGEMENT PERSPECTIVES

The use of BS 7750 has helped companies take another look at the way they do business. Gaining a different perspective on how to increase profits has never been easy and, until the publication of the standard, not many business people would have looked to environmental issues to help increase the efficiency and effectiveness of their organizations.

Yet that is what all the clients of the UK's environmental certification industry have done. They took a fresh angle on the way they manage their business—and they are not alone. They are being joined every day by a growing number of managers who have discovered the positive value of such a system and can link it directly to the profitability of their organizations. If they have a common watchword, it would probably be "Act now to secure the future". But what is this new perspective and how is it linked to the use of BS 7750?

In today's business world, relying on learning from past mistakes is like driving a car while looking through the rear-view mirror. Progress is bound to be uncertain. However, in the present legislative framework, the risks to any organization from the effects of poor environmental management are being realized in measurable financial terms, on a day-to-day basis. Every day, without the benefit of sound environmental management, the cost to any organization can be measured through what are, effectively, invisible financial leaks. It is the hidden drain of poorly used energy; the slow but steady drip of raw materials directly into the waste stream; the needlessly hurried investment in expensive technology to meet ever tightening regulations. A company does not have to be a polluter to pay environmental costs.

The new perspective afforded by users of the standard relates to the way in which they view the complex set of densely packed, interrelated environmental issues that shift and change continuously. The systematic and integrated approach called for by the standard means that managers can now manage business risks in proportion to the threat posed to their company. It allows them to pinpoint the key decisions that help to reconcile flexibility of response with the solidity required by good financial returns. The overall benefits experienced by organizations include gaining tangible financial benefits by plugging the gaps of the past and the ability to express their confidence in their abilities to their market now, thus safeguarding their access to and share of that market in the future.

Users of BS 7750 found that an environmental management system can be used to provide an organization with an understanding of the issues and risks that directly affect their efficient performance. Once they had identified what environmental costs they needed to manage, and to what extent they were managing them already, they used the system to set themselves a series of achievable objectives and targets that:

(a) helped to manage current and future environmental risks;
(b) gave confidence to any party interested in their continued financial and environmental health (banks, insurance companies, shareholders, local communities, local authorities, employees, regulators, etc.);
(c) enabled them to make cost savings through greater efficiency and less waste;
(d) identified future risk areas for further research and evaluation;
(e) allowed them to make improvements at a rate that recognizes changing business circumstances;
(f) gave growing confidence that environmental liability issues have been addressed.

All these aspects of business performance were realized within the context of the organization's existing management systems and culture. This was possible due to the flexible nature of the framework supplied by the standard. No two companies' experiences were alike, and yet all achieved improvements in the way they confronted and managed the environmental impacts of their activities. Case studies within particular trade sectors are already being eagerly sought by those who wish to take up the opportunity offered by the international standard ISO 14001 and, with increasingly positive experiences to relate, these studies will prove invaluable to the environmental managers of the future.

NOTE

Extracts from BS7750:1994 are reproduced with the permission of BSI. Complete editions of the standards can be obtained by post from BSI Customer Services, 389 Chiswick High Road, London W4 4AL.

18
Environmental management system certification—an assessor's view

JEFF DOWSON

INTRODUCTION

Environmental certification began soon after the release of BS 7750 in 1992. As a major certification body, SGS Yarsley saw the need for companies to have their environmental management systems certified by an independent third party as proof of their commitment to the outside world. At this time there was no accreditation service for a certification body to approach to gain accreditation for the work we were undertaking. Several members of Yarsley staff were involved in influencing the standard to allow it to use all the benefits of the quality standards in the ISO 9000 series, and most of all to ensure that the standard was assessable. The early certification was performed without the benefit of the United Kingdom Accreditation Service (UKAS) accreditation criteria, which were developed as the accreditation of certification bodies approached in 1994. The unaccredited certificates were known as "green dove" certificates, after accreditation this name stuck and hence we still know environmental accreditation as green dove, and the dove is our logo for EMS certification.

UKAS developed its environmental accreditation criteria (AU/2/23) as a guidance document to be used in conjunction with the environmental standard and the international standard for the operation of certification bodies, EN 45012. It could not add any new requirements to the environmental standards as this would then be an additional standard, but what it did was to provide the certification bodies and potential clients with a firmer base on which to build their systems, and gave the potential clients the surety that all accredited certification bodies would adopt similar working methods and interpretations of BS 7750 and ISO 14001.

ENVIRONMENTAL MANAGEMENT SYSTEM ASSESSMENT

Accreditation environmental management system assessments take the following route: initial assessment, desk study, main assessment and surveillance visits.

Initial assessment

- To verify that the system was based on significant effects and ensure that it is aimed at controlling and improving environmental performance and is auditable.
- To ensure that reliance can be placed on internal audits.
- To plan for the main assessment.
- To provide immediate feedback to the client which may assist in the remainder of the assessment process.

Desk study

- To ensure compliance with all clauses of the standard and prepare checklists for the main assessment.

Main assessment

- To verify that the system meets the requirements of BS 7750 and/or ISO 14001.
- To ensure that the system is capable of delivering and achieving performance improvement and regulatory compliance.
- To verify compliance with company policy and procedures.

Surveillance visits

- To verify continued compliance and to ensure that the mechanism for continual improvement is working. These visits will occur every six months, with a renewal visit every three years.

EFFECTS EVALUATION

Section 4.4.2 of BS 7750 requires a company to "identify, examine and evaluate" environmental effects (ISO 14001 refers to environmental aspects, clause 4.3.1). The standard then lists requirements to consider:

- emissions to air
- discharges to water
- solid and other wastes
- contamination of land
- resource usage
- noise, dust, vibration and visual impact
- effects on parts of eco-systems

Of these effects, most companies do not have any problems with air, water and land, as these can be easily seen or accounted for by a simple process diagram backed up by chemical equations. Resource usage also seldom causes too many problems, as this can be evaluated by looking at raw materials, energy and water consumption. The three commonly recurring areas that are seen as problem areas on assessment are evaluation of:

- contaminated land
- noise, odour, dust, vibration and visual impact
- effects of eco-systems

If you add to that the fact that environmental effects should be considered according to the following criteria:

- normal operation
- abnormal operation
- incidents, accidents and emergencies
- past activities

the problems are made even greater.

Contaminated land

Contaminated land is a topic that provokes intense discussion, especially when considered as a past activity. With changing legislation and the fact that the current owner is responsible for the land, the topic is now addressed a little better but the attitude of "let sleeping dogs lie" usually prevails. Companies seem reticent to look to hard at contamination of land in case they are faced with hefty clean-up bills. In cases where the site was previously "green-field" the problems are easily solved, but the environmental effect still needs to be documented. Often clients are unsure what they should be doing to identify any problems. Where sites have changed hands several times and the processes have changed there is often little record of the potential contaminated land as the exact usage of the land may be unclear. The best approach to this topic seems to be to do the best possible rather than nothing.

Noise, dust, vibration and visual impact

As so many organizations in the UK are familiar with the requirements of the Health and Safety at Work Act and Control of Substances Hazardous to Health (COSHH), it is strange that noise, dust etc. are so often missed out. Companies keep registers of correspondence showing complaints of noise and dust etc. but these topics then fail to make the environmental effects register. It is usually under abnormal operating conditions that these environmental issues arise, e.g. during repairs or building construction and maintenance.

Effects on eco-systems

Effects on the eco-system are so often misunderstood and thus left out of the environmental effects evaluation. In essence the effects referred to are those on other life forms resulting from the operation of the process at a particular site, for example dust can cover plant leaves reducing their effectiveness in photosynthesis and thus the plant's growth is stunted or in extreme circumstances it will die.

Direct and indirect effects

Direct effects of the operation are usually well covered, but companies seem unsure of how far they should go when considering their operation's indirect effects. Some companies are even unsure of what indirect effects are. An indirect effect is one which occurs as a result of a supplier or end user's actions, e.g. extraction of raw materials or disposal of products after use. Another common problem with indirect effects is how far up and down the supply chain to go. It is not always necessary to go from cradle to grave, but good advice is to look at least two steps up and two down the supply chain. By doing this the company should avoid unfair criticism of its evaluation.

For any environmental effect the company must be prepared to justify the application of its own procedures. The auditors will question some effects and look for consistent application of the environmental effect evaluation procedure.

Defining significance

BS 7750 calls for the company to classify its effects according to significance. Assessors are presented with many ways of attaching significance and it would be hard to say which method is correct as each case must be considered on its own merits. Some companies have used a mathematical scoring system to define significance. The scoring systems are weighted in the favour of non-compliance, with legislation being the most penalizing. These systems work well until the accident and emergency situation is considered and then confusion appears about the level of significance. Other definitions of significance have involved the use of environmental reasoning where each effect is studied for its potential impact on the environment and the level of significance is defined by a panel of people within the company. There is no right or wrong way to define significance: all that is required is for the company to be able to justify its reasoning to the audit team and show that the rationale has been applied consistently.

Not all environmental effects are detrimental ones. Some companies neglect to show beneficial effects. For example, beneficial effects can include a water treatment plant that returns water to a water course cleaner than when it was removed and thus regenerates the water course.

THE MANAGEMENT SYSTEM

The accreditation criteria require significant environmental effects to be managed. The management can take two routes:

1. By operational control to ensure continual compliance and maintenance of performance.
2. By setting an improvement programme, e.g. by setting objectives and targets.

Operational control

Operational control requires the setting of controls which must be measured for performance to ensure that they are effective. Figure 18.1 shows the loop required.

An operational control is set to ensure that the environmental effect is controlled. The control is verified by measurement. If it is out of specification, corrective action is taken to bring the effect into control. Then the loop is revisited.

Management programme

An improvement programme requires the significant effect to be allocated an objective for reduction in its effect and a target for the completion of the reduction.

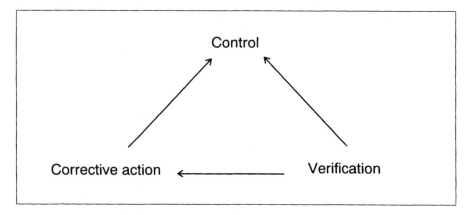

Figure 18.1 Operational control loop

POLICY INTEGRATION

Companies that are part of a larger group often find problems with integrating their environmental policies with those of the parent body or group. Section 4.2 of BS 7750 states that the policy should be defined "within the context of any broader corporate body". This statement, although simple in meaning, can cause many problems. Some sites of large multinationals can find themselves with three or four higher-level policies with which they have to comply, and sometimes these higher-level policies mention objectives that are considered trivial by the site concerned. The standard requires that all higher-level policy requirements are referenced in the policy of the site. It should also be noted that in annex A.2 of BS 7750 it says that the environmental policy statement should be consistent with the occupational health and safety policy and other policies (such as the quality policy).

In practice, the specific BS 7750 requirement has been achieved by the site referencing related documents in a company's policy, holding controlled copies of the other policies, and being able to demonstrate that when one of the related higher level policies changes these changes are reflected in the site policy.

REGISTER OF LEGISLATION

Another common area of concern is the register of legislation. In the UK there are over 600 pieces of national and European Union legislation that can relate to a company's activities. A company's managers should consider which legislation is important to them, because by not fully understanding the legislation that applies to the company they could be breaking the law. In addition, managers should have a method of ensuring that they are aware of any proposed changes to legislation which may affect their organization, as this could have an impact on improvement plans, objectives and targets. Thus to gain the most from the register of legislation it is suggested that the register be produced as a very focused document, rather than just a long list of Acts of Parliament. Better registers of legislation not only list the legislation but also present a précis of how the legislation affects the company.

Should the environmental management system be solely based on legislation? Definitely not.

The company would not derive any benefit except legislative compliance from an environmental management system based on legislation alone. In addition, in the unlikely event that the legislation applicable to the organization does not change, the EMS would not be able to show continual improvement and thus would not meet the requirements of BS 7750 and ISO 14001. If the EMS was established in an organization in a country with undeveloped legislation the EMS would be achieving little.

The EMS should be based on the management of those environmental effects that the company has decided are significant, and by this the maximum benefits will be gained.

OBJECTIVES AND TARGETS

Objectives and targets are set as part of the improvement programme. Their aim is to quantify what improvement is to be made over what period. Often objectives are presented with an "ongoing" status. Ongoing is not a quantifiable time scale and thus the performance cannot be measured. Companies should try to break targets down into smaller, more manageable pieces. For example, the following objective is common:

Target	Time scale
Assess all environmental performance of suppliers	Ongoing

A far more measurable approach is:

Target	Time scale
Send environmental questionnaire to suppliers and ask for policy statement	Month 1
Ensure that all questionnaires are returned	Month 2
Evaluate responses	Month 3
Contact those with significant impacts	Month 4
Review impacts	Month 5

etc.

In the latter model, managers will be able to measure performance after a month as they will be able to see if the questionnaires have been sent out. In the first case after one year there may not have been any action as no definite target was set.

If an objective is not achieved within a targeted time scale, a third-party assessor will not always raise corrective action against the system. However, if there has been no corrective action put in place to correct the deficiency, then this will mean that the system has broken down. In these circumstances assessors will always raise corrective action and this will be considered as a serious area of concern.

INTERNAL AUDITS

Often there is the misconception that if there are quality auditors on site then they will be able to audit the environmental management system without any further training. This is totally incorrect. Good environmental auditors will have the following four skills.

1. They will be trained to audit management systems.
2. They will have knowledge of the environmental management systems standard being used, e.g. BS 7750, ISO 14001 or EMAS.
3. They will have knowledge of the process being audited.
4. They will have suitable environmental knowledge to assess the impacts of the effects being assessed.

Quality auditors will only have skill number 1 out of the four listed. This does not mean that quality auditors are of no use in auditing environmental management systems, but that such auditors will require more skills training to make them useful. It has been noted in some companies that the independence of some auditors has been sacrificed at the expense of ensuring that the audits were carried out by persons possessing all of the four skills listed above. As environmental management systems become more common this independence will develop since more suitable auditors will be trained.

During the assessment process the internal auditors are usually questioned by the external assessors to ensure that they possess the four skills, because the assessor is looking to ensure that the internal audits are reliable at highlighting internal problems.

Internal audit schedule

As with all other aspects of BS 7750/ISO 14001, the audit programme must reflect the significance of effects which have been defined. A common problem seen with EMS audit schedules is that the audit schedule emulates an ISO 9000 schedule where each activity is given equal importance, i.e. everything is audited twice per year. The EMS audit schedule may be designed in such a way that some areas may be audited only once every two to three years because the area had minimal significance whereas key issues could be audited two or three times per year.

BS 7750 states in section 4.3.4 that training should be given to persons whose work may have a significant effect on the environment. In practice, it would be expected that process operators would know what controls were being applied to the process and why. At a higher level it would be expected that the manager of the process would have knowledge of the environmental effects of the process and the effect of non-compliance. It is not acceptable, as with standards in the ISO 9000 series, for the training of all staff to be simple awareness training, as this would not reflect the importance given to those activities seen as significant.

INTEGRATION OF MANAGEMENT SYSTEMS

The accreditation criteria (AU2/23) promote the idea of integration of management systems. It would be to a company's advantage to have only one management system for all of its activities, e.g. quality, environment, health and safety. The problem is smooth integration. There are some complications, however, ISO 9000 series standards will often promote just-in-time (JIT) methods of production, whereas BS 7750 or ISO 14001 may oppose this as not being the best environmental option. Nevertheless, once such problems are overcome one system will be better than two or three, as a common theme can then be applied to all areas and activities.

CONCLUSION

EMS certification is still in its infancy. BS 7750 was first issued in 1992 and ISO 14001 in 1996 thus interpretation is not as fixed as with quality standards such as those in the ISO 9000 series. Consultants have not proliferated in the EMS field and so "off-the-shelf" systems have not been developed.

When a company is preparing an EMS it should bear in mind that the systems should be implemented to help the company and not merely to comply with the standard. A good EMS combines with the quality system to provide one overall company system.

19

Certification and harmonization of environmental management systems

ROGER BROCKWAY

INTRODUCTION

This chapter examines the role of the certification of environmental management systems (EMS). I do this from the point of view of a national accreditation system. The functions of national accreditation are to ensure that independent certification can be trusted by the market and, through harmonization internationally, that certificates are understood and accepted globally.

ENVIRONMENTAL MANAGEMENT SYSTEMS

Every company has some form of EMS. In the past, many such systems have been an unconscious part of the organization and therefore not managed to optimum effect. Standards—in particular the International Standard ISO 14001—now set out the necessary elements of an EMS. Companies can use these standards to give full definition to their own systems and to maximize the potential of these.

This potential has two interrelating elements:

- internal efficiency
- external performance

Environmental Management Systems and Cleaner Production, edited by R. Hillary.
© 1997 John Wiley & Sons Ltd.

Companies with a good EMS will therefore enjoy the joint benefits of:

- improved internal efficiency, derived from taking positive control of the management of the environmental aspects of their business; and
- improved environmental performance through the minimization of the harmful and maximization of the beneficial environmental effects of their activities.

The whole economy benefits from the former. The planet benefits from the latter. I believe that there is much to play for and my theme in this chapter is to give confidence to the market that these benefits are achievable through accredited certification of EMS.

ACCREDITED CERTIFICATION

What contribution does accredited certification make to confidence building in the market?

EMS standards may be beneficially applied by any business, and self-declaration of compliance is an acceptable practice. However, certification (or registration) by independent third parties that the EMS does meet the requirements of the standard is normally a requirement of those economic partners and stakeholders in the business who have an interest in its system. These economic partners and stakeholders do not normally have the means to check the compliance of the system, nor would it be a good use of resources for each of them to do so. Independent certification also helps to ensure that the system is maintained and responds to changes in the company. Without it, there is a real danger of management commitment to the system being downgraded in response to other pressures.

Third-party certification gives confidence. To do this it must be:

- independent
- competent
- consistent

It is the job of accreditation to ensure that third-party certification meets these three requirements. To do this, accreditation bodies have criteria, based on international standards, that they apply to the certification bodies. Once accredited, the certification bodies are subject to regular (six-monthly) surveillance.

Independence

Independence is assured through checking that the structure of the certification body ensures that decisions on certification will be taken impartially by people without any special interest in the result. Investigation is also made to ensure that any potential for conflicts of interest affecting all parties in the certification process, including auditing, will be revealed, analysed and managed so as to be effectively neutralized.

CERTIFICATION AND HARMONIZATION OF EMS 185

The same independence requirements can apply to certification bodies acting in different fields.

Competence

The latter comment is not true of competence. To certify compliance with EMS standards, certification bodies must demonstrate possession of relevant specialist competencies. These are:

- understanding of the standard;
- understanding of environmental protection;
- technical knowledge of the activity to be certified;
- knowledge of environmental legislation relating to the activity;
- management system assessment skills.

To be accredited, a certification body must be able to demonstrate its possession of these competencies at two levels: first within the management of the certification body, and second in its audit personnel.

Head office competence

The certification body's management needs the competence to understand the environmental aspects of the business of any organization that it undertakes to certify. It must have the competence to:

- appoint the right audit team; and
- take a decision on certification based on the audit team's report.

The first is done through a competence analysis of its client needs which includes the following questions:

- What are the environmental effects of the client's activities?
- What skills are needed to recognize these effects in operation when performing an audit of the company?
- What audit personnel are available to the certification body who have, between them, the necessary skills?

When certification bodies are accredited, they are only accredited to work in defined scopes. Their scope limitation will be tied to the areas where they can demonstrate ability to perform competence analysis along these lines and availability of human resources.

Audit competence

Checking the competence of the audit personnel in the certification body is at the heart of the accreditation process and is done through witnessing audit teams in action. At this stage in the development of EMS certification it is unusual for a single person to be able to conduct a certification audit, hence the emphasis given to selecting appropriate audit teams. UKAS cannot rely on any single professional qualification to assure it that audit personnel will be competent, although environmental auditor registration schemes will become increasingly useful with experience. However, auditing for EMS certification is a specific part of the environmental auditing profession that the accreditation bodies will have to oversee with care.

The standards in the ISO 14000 series relating to environmental auditing embrace requirements for first-party (internal) auditors, second-party (consulting) auditors and third-party (independent) auditors. One of the skills of the third-party auditor is to know how much reliance to place on the internal audit, and not to duplicate it unnecessarily. There is a useful commonality in the requirements for these three types of auditing activities, but more important to establishing competence for third-party EMS auditing is rating against the five competence areas set out above.

For a certification body to work effectively, it will mix and match audit personnel with different backgrounds and complementary skills. What qualifies an audit team is, collectively, understanding how the company's EMS achieves its aims, regulatory compliance and the measures of performance improvement that derive from the company's environmental policy. The team must demonstrate its ability to track the company's EMS through the following stages:

- environmental effects analysis
- evaluation of significance
- management of significant effects
- monitoring of performance
- corrective action

To do this, the team must have knowledge of the industry, of the regulations under which it works, of its technological range and performance parameters. This does not mean that the team should try to impose its own ideas on the company of what the EMS should be achieving. Rather, it should have the skills and knowledge to recognize the actual environmental performance being delivered by the company's EMS. If failures are found in performance delivery these cannot, in themselves, be translated into non-compliance in the EMS. A good audit team will track such failures back into the EMS to discover any underlying failure in the system. This may, for example, reveal a basic failure in the system for evaluating environmental effects, resulting in the important effects not being addressed and managed by the system. The audit team needs the skill and understanding to bring these things to light and to track them down to the underlying failure in the system.

Thus performance improvement, which is the aim of the EMS, provides the focus that the audit team uses for reviewing the effectiveness of the system. Persistent non-achievement of planned performance improvement must have a root cause. The good audit team will find this, adding value to the certification process.

This means that EMS audit teams must be technically at home at the company's site and concentrate on results in the field rather than on documentation of the system.

ISO 14001/ISO 9000

This essentially dynamic aspect of EMS auditing is, quite logically, in contrast to the way in which quality systems to the ISO 9000 series of standards have often been audited in the past. The ISO 9000 series describes management systems designed to meet customer requirements. Its focus, therefore, is on compliance. ISO 14001 describes a management system which is not driven by a customer and therefore has itself to contain the elements that deliver performance. Certification therefore focuses on the actual performance. The system, of course, has to be documented and the audit team needs to check this, but it is performance, not the documentation, that shapes the audit.

This, as well as the obvious difference in competence requirements, is a major reason that traditional ISO 9000 certification bodies have found it hard work adapting to EMS certification. It is also a reason to be very careful in introducing single certification of management systems that integrate environment with quality and other aspects. The market demands such a service, and accredited certification will work with the development of standards to meet it, but with caution so that the essential elements of each aspect are not lost.

PROGRESS IN UKAS ACCREDITATION

UKAS, the United Kingdom Accreditation Service (formerly NACCB), has had the task of accrediting EMS certification bodies in the UK since the end of 1993. In 1994 UKAS conducted a pilot exercise based on the UK EMS standard BS 7750. This led to the development of accreditation criteria, against which, from 8 March 1995 till October 1996, UKAS accredited 17 certification bodies. Fifteen of these are UK based. Before 1995, UKAS was only allowed by its statutes to accredit within the UK. However, it is now allowed to accredit anywhere in the world and has already accredited bodies in Japan and the USA.

Now that ISO 14001 has received international standard status, UKAS is authorizing accredited certification bodies to give certificates of conformity with this full international standard which has been adopted as the British Standard BS EN ISO 14001 (BS 7750 will be withdrawn on March 31st, 1997). UKAS is also appointed in the UK to accredit environmental verifiers under the European Union's Eco-management and Audit Scheme (EMAS).

HARMONIZATION OF ACCREDITATION WORLDWIDE

Accreditation operates primarily at the national level. If EMS are to be certified consistently across the world it is necessary for the national accreditation systems to harmonize the criteria used for accrediting certification bodies. It is also necessary to look at any differing standards used in parts of the world, such as EMAS in the European Union, and to make clear to the market what the differences between the standards are. UKAS aims to ensure that work done by a company on its environmental system will receive as wide a recognition through one act of certification as possible. Work done for ISO 14001 therefore receives recognition for all equivalent aspects of EMAS. This is because one accreditation system provides assurance that the common elements are identified and implemented at a satisfactory level.

The national accreditation systems in Europe have been working together on environmental accreditation in the European Accreditation of Certification (EAC) since 1993. The UKAS criteria for BS 7750 certification have been taken and developed into EAC criteria, applicable to all forms of EMS certification. The criteria were adopted in 1996 in the 15 European Union member states and a mechanism for peer review between the accreditation systems will be set up to ensure that all national accreditation systems implement the criteria in a similar fashion. This will provide for a single accreditation system, maintaining common standards, that can be entered at a number of points and in a number of languages. It is hoped that the results will be accepted by the European Commission and EU member states as providing for certification procedures recognized under Article 12 of the EMAS regulation, thus ensuring that EAC-accredited certification of an EMS to ISO 14001 will be acceptable as a basis for compliance with EMAS.

The next task is to globalize this principle. An International Accreditation Forum (IAF) already exists and has appointed a task force to develop accreditation criteria for certification to ISO 14001 during 1996. The work of EAC will contribute to this initiative. The intention is to use IAF to bring together all groups of regional accreditation systems to provide a single, worldwide, multi-entry, accreditation system to oversee uniform certification and implementation of ISO 14001 and any other EMS standards.

CONCLUSION

The intention of this chapter is to demonstrate that certification of EMS is not a passive, bureaucratic activity. Accreditation aims to ensure that certification adds value and assists companies across the world to manage the environmental aspects of their business in a dynamic manner, and to use auditing as a tool in this process.

The result should be improvements in environmental performance that will satisfy shareholders while at the same time making our planet a better place to live.

Section IV
European Industrial Experience

20
Introduction

RUTH HILLARY

This section of *Environmental Management Systems and Cleaner Production* is one of three sections which cover the practical industrial experience of enterprises implementing environmental management systems and cleaner production projects. European industrial experience is considered from the European Union countries of the UK, Sweden, Denmark and Norway and from the oil, electricity-generation, manufacturing, chemicals, food-processing and environmental-consulting sectors. Chapters 21 and 22 analyse the integration of quality, environmental and health and safety management systems; Chapter 23 discusses risk-assessment techniques applied to environmental management; Chapter 24 analyses the evaluation of environmental performance in industry, and the final chapter considers the realization of cleaner production through environmental management systems. The central theme of this section is how organizations use an increasingly broad range of tools and techniques to move towards integrated management systems as a means of controlling their environmental performance.

In Chapter 21, Johan Thoresen discusses the potential for environmental performance improvements through the implementation of integrated management systems, illustrating his discussion with Scandinavian case studies from, among others, the oil and manufacturing sectors. Thoresen argues that the systematic use of management systems should complement, and may substitute for, technological investments aimed at reducing environmental impacts from industry. He suggests the systematic use of integrated management systems, product life cycle assessment principles and pollution prevention as means to enable companies to realize the opportunities and cope with the threats and uncertainties arising from environmental issues. Thoresen concludes that companies need to address environmental issues at a strategic level and seek to control them by implementing a total management system which engages all employees.

Stuart Aaron, in Chapter 22, continues the theme of integrated systems by providing an insight into the UK Chemical Industries Association's (CIA) Responsible Care programme, which has the objective of improving the performance and reputation of the chemical industry by providing guidance on integrated health, safety and environmental management systems. He points out that Responsible Care guidance is designed to address the international quality standards series ISO 9000, the British environmental management system (EMS) standard BS 7750, the Eco-management and Audit Scheme (EMAS), the international EMS standard ISO 14001 and successful health and safety management. He describes chief executive officer (CEO) commitment as the key driver of improvements. The CEO is placed at the top of a management cycle based loosely on Deming's "plan–do–act–check" cycle, and Aaron concludes by asserting that the CIA strongly advocates the concept of integrated management systems.

Often potentially higher-risk sectors such as the chemical industry described by Aaron have linked health, safety and the environment and employ the risk-assessment techniques utilized in health and safety to the environment field. In Chapter 23, Paul Pritchard introduces the concept of environmental risk-based management, arguing that the risk-based approach has the twin benefits of being consistent with both business thinking and the nature of environmental impacts. He asserts that formalized environmental management systems such as ISO 14001 and BS 7750 seek to identify the environmental risks facing organizations and describes a number of approaches and tools, such as risk transfer and environmental audits, utilized by UK companies to assess their environmental risks and opportunities. Pritchard concludes that formal environmental management systems are very useful in achieving improved environmental performance, but that the improvements achieved may be subjectively assessed or have significant associated uncertainty.

In Chapter 24, Eskild Holm Nielsen, focuses on the issues of assessing environmental improvements in industry raised by Pritchard by describing the evaluation and quantification of environmental improvements from cleaner production pilot projects in the Danish fish-processing industry. He analyses the reduction in the degree of environmental pollution from the pilot companies by assessing their reduction in water consumption and discharges of organic matter (as measured by chemical oxygen demand, COD). He shows that collectively the pilot companies significantly reduced their water consumption and halved their COD discharge. Nielsen argues that the motivation for companies to adopt cleaner production projects arose because they had become more environmentally aware and because of the threat of future tighter legislation. He concludes that for cleaner production projects to have greater effectiveness they need to be linked to environmental management systems.

Nils Thorsen, in the final chapter in this section, also considers industrial experience in Denmark. He considers the realization of cleaner production and process innovation through environmental management systems implementation, arguing that the Achilles' heel of such systems is their inability to guarantee the choose of cleaner/best technology required by Danish legislation. He considers the economically viable application of best available technology (EVABAT) in EMAS and the question of whether or not cleaner production is a competitive issue. Thorsen cites industrial case studies from the chemical, manufacturing and consulting sectors. In conclusion, he asserts that if environmental management systems are to live up to current ecological challenges facing industry they must be able to promote cleaner production in a significant way.

This section's European industrial case studies illustrate how enterprises go through a number of stages in environmental performance management. These stages may not run consecutively, but industrial experience shows that an enterprise usually starts with technological controls, typically end-of-pipe, and that these controls are gradually substituted by clean technologies. At the same time, management systems are established, often for separate functions, starting with health and safety and then quality, which may be formalized and certified to standards such as those in the ISO 9000 series. Techniques and tools such as risk assessment and environmental auditing are utilized throughout these stages. Environmental management is then introduced, sometimes as formalized systems such as BS 7750, EMAS and/or ISO 14001. Mature enterprises seek to integrate their systems and bring them into one holistic management system (see Figure 20.1). Once this has been achieved or is well established, an enterprise begins to look outside its boundaries to its product life cycle and stakeholder values.

This section addresses a number of additional and reccurring themes:

- The need to link cleaner production to management systems to make the benefits sustainable.
- The role of environmental management systems in complementing, promoting or even substituting for technological investment.
- Companies' innovation of and access to a wider range of tools, including life cycle assessment, risk-assessment techniques and process-innovation procedures, to assess their environmental impacts.
- Systems integration as a key long-term aim and achievable objective of enterprises seeking to manage their environmental performance.
- The need for enterprises to think strategically about environmental performance and to link this performance to business objectives and competitiveness.

The high degree of integration and sophistication of European enterprises appears mainly to be concentrated in larger, transnational organizations rather than smaller

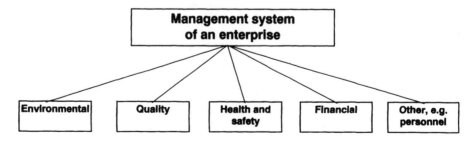

Figure 20.1 Holistic management system of an enterprise

national enterprises. The practical experience of these larger companies contrasts dramatically with the case studies presented in the following section, which draws on industrial experience from emerging and transition economies and the challenges these industries face on entering the international market.

21
Potential for improving environmental performance through implementation of integrated management systems

JOHAN THORESEN

LIMITATIONS OF TECHNOLOGY AS A SOLE SOLUTION TO ENVIRONMENTAL PROBLEMS

An industrial company is an open system in the sense that it is continuously influenced by external factors signalling rapid or longer-term change, e.g. changes in customer preferences, changes of political instruments used for control, shortage of non-renewable raw materials, technology change. Such factors represent future threats, future uncertainties and future potential for company development. Investment decisions in process or end-of-pipe technology to minimize environmental impacts are often based on a static view of product design or specifications, with limited reference to company-external factors other than the environmental authorities. End-of-pipe or process technology investments may therefore prove not to be optimal as the sole answer to handling environmental problems. Company decisions, development activities and operations should therefore be closely linked through an integrated management system, securing a holistic view of the problems to be solved

Environmental Management Systems and Cleaner Production, edited by R. Hillary.
© 1997 John Wiley & Sons Ltd.

and interconnection between the various company activities. This is illustrated by the two following cases.

An integrated management system is in this chapter broadly defined as the principles, structures, procedures, decision-support tools and human processes at all organizational levels that are necessary to carry out strategic and operative management. The overall objective of the management system is to secure short- and long-term profitability and survival.

In large organizations, development engineers are normally engaged in solving technical problems within a limited area of a total development project, and have too little information available to fully understand the extent and complexity of the total project. A holistic approach to solving the total project to reach the economic/technological/operative/environmental optimum solution should therefore be the implementation of an integrated management system. Case 1 (Thoresen 1991) shows that consequences of relevant, environmental and operative issues should be continuously integrated into project decisions during the development period to reach the best possible trade-off between project economy and environmental and operative requirements.

Case 1   Norwegian off-shore investments

> When Saga Petroleum commissioned its tension leg platform in the Snorrefield in 1992, it was the successful result of a complex planning process stretching over nine years. Off-shore technology has advanced extremely rapidly during the last 10 years, to being cheaper and more efficient. During the same period, the Norwegian authorities became aware of the negative effects from North Sea oil production on the regional and global environment. Therefore increasingly stricter regulations and taxation of emissions were introduced, e.g. the CO_2 charge amounting to some £23 per tonne of CO_2 for offshore production installations.
>
> For economic reasons, changes in platform concept in this kind of project will not be accepted later than the start of the detail engineering phase, in the Snorre case 4.5 years prior to commissioning of the platform. A study had shown that by changing the gas-turbine concept on the platform, a 33% reduction of CO_2 and other combustion gases was possible. Furthermore, the net present value (NPV) of the reduction in future CO_2 charge would exceed the NPV of the necessary investments by some £1 million, provided that this investment decision was made prior to the detail engineering phase. A decision delay until after the starting point of detail engineering would increase the necessary investment cost by a ratio of 25:1.
>
> *(continued)*

(continued)

> This case clearly shows that a continuous flow of important information to the engineering department in connection with important decision milestones—in this case, an update on the approaching CO_2 charge—is essential to reach the most cost-efficient total solution. This can be secured through various parts of a management system.

Experience from a large number of Norwegian industrial companies during the last five years has shown that very large environmental improvements, with no or very limited cost consequences, have been attained purely by putting the management focus on environmental problems and by using a systematic way of planning and implementing internal changes (see Case 2; SFT 1993). This indicates the importance of a management system which is based on a continuous and systematic way of defining, assessing and attacking pollution problems, where a first priority should be to implement no-cost or low-cost improvements to eliminate or strongly reduce important sources of pollution. The second priority—if necessary—should be the installation of relevant and cost-efficient technology.

Case 2 A Norwegian programme for pollution prevention

> Statens Forurensningstilsyn—the Norwegian Inspectorate for Pollution Control—during the period 1991–93 financially supported the introduction and use of environmental assessments in 150 Norwegian industrial companies. By 1994, 48 companies had completed their analyses and improvement assessments, concluding that it is normally possible to obtain reduction in emissions to air, releases to water and production of waste amounting to between 20 and 50%. Typically, the payback period for the necessary improvement investments was shown to be below two years, and 77% of the companies reported that a considerable environmental improvement potential had been discovered where no process investments were required. These improvements gave economic benefits and were mostly attained through "no-cost" changes of internal procedures and instructions.

INTEGRATION OF ENVIRONMENTAL ISSUES INTO SYSTEMS FOR STRATEGIC AND OPERATIONAL MANAGEMENT

A loop for strategic management is suggested in Figure 21.1. Clarification of the "rationale" for the company or its "business idea" precedes the development of company policies in important areas, for example environmental policy, policy on company ethics etc. Against this backdrop, company objectives are developed before the important strategic analysis of external and internal factors with a bearing on future company profitability is carried out. In this context, company strengths and weaknesses are mapped out to define what changes are necessary in the company when it decides to take advantage of external opportunities and to enable it to handle threats or uncertainties in its environment. The next phases are the setting of strategic goals and deciding which strategies will satisfy those goals. This leads to the choice of strategies regarding six different strategic aspects that need to be addressed: marketing, operations, human resources, finance, information technology (Thompson 1993) and environmental impacts from products and processes. Normally these aspects are discussed in parallel to secure integrated strategic treatment before the subsequent strategic choices are made.

The choice of strategies leads to the definition of relevant strategic programmes and activities (Johnson and Scholes 1993). The implementation of the programme

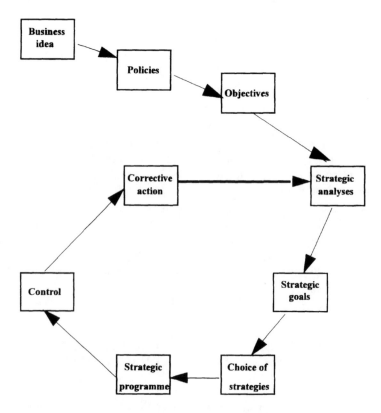

Figure 21.1 Loop for strategic management in industrial companies

will take place at departmental level. Results are monitored by top management to make sure that targets are met and, if not, that corrective action is initiated. Environmental aspects are integrated into this process together with individual strategies, targets and programmes in combination with the other strategic aspects in a common, company-wide strategic programme.

It is imperative that the strategic process is not a bureaucratic exercise but a planning process that enables managers and key personnel to understand which issues are important and how they should be addressed. In many companies the strategic planning loop is run every one or two years. The frequency of this fundamental and company-wide process should be tailored to the needs of the individual company. Many companies find themselves in the middle of a dynamic and changing environment that demands continuous assessment of external factors and rapid adjustment to them. Consequently a more proactive style is imperative, where the strategic planning is made continuous, creative and with no fixed frequency. The individual strategic aspects are addressed according to their importance and rate of change when influenced by the company environment (Clutterbuck 1991).

Operational management makes things happen according to plans and procedures on a day-to-day basis inside the company. Since there is a complex set of activities that has to be carried out in an industrial company, the plans, procedures and follow-up activities have to be formalized as part of a total management system. A strategic management system, a quality management system or an environmental management system are all examples of elements of an integrated management system.

The development of the ISO 9000 series for quality management, the Eco-management and Audit Scheme (EMAS) introduced by the European Union in 1993, the British environmental management system standard BS 7750 and the future ISO 14000 series to be introduced in 1996 are all examples of ways of standardising routine working methods for operational management. A requirement for a policy statement and the setting of objectives for future improvement connects these systems to strategic management. It is important to note that the above systems only describe a structure for how the company's internal improvement activities should be carried out. The success of such systems rests with top managers' ability to implement the system in the organization by involving and motivating their staff.

One factor often overlooked, but with a strong impact on successful systems implementation, is to what extent company employees—managers as well as other employees—possess the necessary competence and motivation required for their jobs and responsibilities. Thoresen (1994) points out that it is essential for company managers to understand and take advantage of this. At the same time top management must be able to initiate activities (Beer and Walton 1990) or adjust working conditions so that the major part of the organization is motivated to perform to their best competence and ability in their jobs and perform as an effective part of the total organization.

INTEGRATED MANAGEMENT AND LIFE CYCLE PRINCIPLES DEMAND SYSTEMS THINKING IN TWO DIRECTIONS

To gain a complete understanding of which external and internal forces influence an organization, this chapter chooses to look at the organization from a systems point of view (Senge 1992). A management system needs to be based on a holistic analysis of external and internal forces acting on company management and to make clear what the impacts are from individual forces—or the interplay between individual forces—on decision making, initiation and control of activities. According to Asbjørnsen (1992), it is of major importance to define the correct border around the system to be analysed.

The author views the integrated management system as consisting of two interconnected systems. The sequence of the different company decisions and levels of responsibility stretches out along a vertical axis. The sequence of cradle-to-grave issues constitutes the horizontal axis (see Figure 21.2). In the vertical direction, it is essential to base management decisions and actions on the integrated management system. In order to develop a system that will meet company objectives, the interaction of top management with its environment (external factors) and with the rest of the company will need to be considered.

Decisions are normally based on a complex set of data and information and require a functioning management system to become optimal.

The second system border encompasses important life cycle issues concerning the company products, i.e. from raw material extraction to end use and waste handling. Then environmental impacts from manufacture, use and disposal of the products are all included. The traditional way of improving environmental impacts from manufacturing processes has been through process and end-of-pipe investments. Normally manufacturers have been forced to act quickly to improve their environmental performance and environmental regulation or charges have been the constant threats for not acting. Traditional end-of-pipe investments in their own manufacturing processes have usually been the result.

The life cycle profile in Figure 21.3 points out that this type of action may be very wrong. This figure shows the contribution to regional environmental problems from the manufacture, use and disposal of plastic bags, exemplified by the acidification impacts. The profile not only points out the location and relative importance of contributions from the individual manufacturing processes along the product life cycle, but also the relative importance of the various transportation/distribution processes and energy demand.

The profile clearly illustrates that, before the manufacturer decides to invest in process improvements or end-of-pipe solutions, it is imperative to define where in the life cycle the environmental problem is largest and where improvement actions can be implemented most cost-effectively. Management systems with tools for analysis and decision support are therefore necessary as backdrops for individual investment

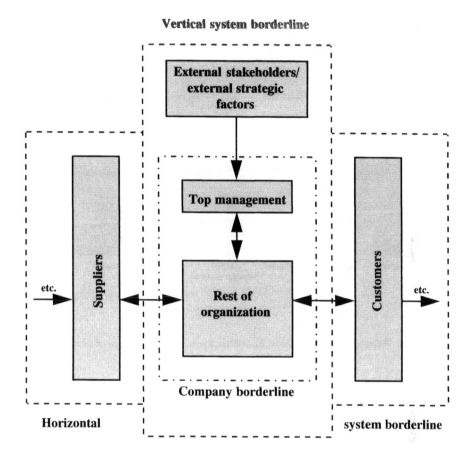

Figure 21.2 Two-directional systems thinking on company priorities

decisions. In this way it will be possible to define the correct priority for improvement action and possible cooperation with important actors along the product chain (Blanchard and Fabrycky 1989).

PREVENTION RATHER THAN PROCESS IMPROVEMENTS AND END-OF-PIPE SOLUTIONS

According to the EU's Fifth Environmental Programme (Fleming 1993) which will be the European Union environmental policy guide for the next decade, prevention has been given a top priority according to the following list:

- prevention
- reuse or recycling

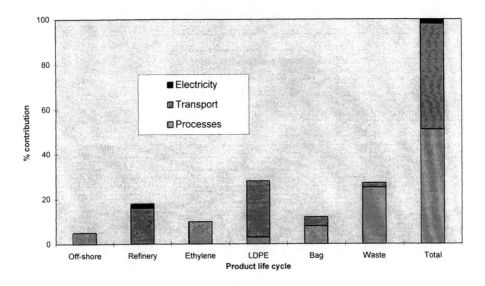

Figure 21.3 Percentage contribution to acidification in the life cycle of plastic bags

- combustion as fuel
- disposal by incineration
- disposal in landfill

When a company keeps to the traditional way of environmental improvement through process and end-of-pipe improvements, it at the same time tends to regard its products in a very static way. By prioritizing pollution prevention at or close to the source, a new form of creativity may be introduced into the company, where a product may be looked at more freely as a means of satisfying specified customer needs rather than being defined by the available production processes and current product specifications. The logical step therefore is to cooperate with the customer during the product development stage (Hanssen 1994)—by using quality function deployment—to arrive at product specifications that emphasize customer needs, quality, cost and environmental considerations. This may imply the use, for example, of alternative raw materials, new and more environmentally-friendly suppliers or completely new designs or specifications. A company policy and management system that address the principles of product development and customers and supplier involvement is essential to force internal working habits and procedures in the desired direction. Case 3 (Amundsen 1993) shows the potential for such changes of habit.

Case 3 Improving environmental performance through changes in product design and raw material base

> Skrovbyggarna AB is a small Swedish company producing sail and outboard engine-driven leisure crafts in glassfibre reinforced polyester. Strict demands from the pollution authorities to reduce styrene emissions from the polyester into the working atmosphere forced the company to review its process and product design. The price for the necessary end-of-pipe cleaning equipment proved to be prohibitive. Modifications in polyester application techniques and a change to a polyester material with better environmental performance showed the potential to reduce styrene emissions by 73.2%. A new hull design based on high-quality fibre reduced the demand for polyester and lowered hull weight by 35%, also producing another 14.9% reduction of styrene emissions. Although the new raw material and application techniques increased the cost of the hull, the company buying the hulls for outfitting and finishing regarded the new hull as having superior performance due its lower weight and improved strength. Increasing the price to the end customer was not therefore regarded as an obstacle. The systematic analyses of improvement potential therefore showed that preventive measures can produce a significant environmental improvement for a product where the price of the necessary end-of-pipe technology proved to be prohibitive.

The lesson is that preventive measures introduced by improved design or specifications in the product itself, combined with improvements triggered by supplier and customer cooperation and assessment, may have a very significant impact on environmental performance. Such measures may substitute for or complement investments in process and end-of-pipe technology, but require identification via a structured decision and control system, normally supplied by a management system.

CONCLUSIONS

A company must be able to cope with the opportunities, threats and uncertainties presented by external factors, and this is made possible by developing and using an integrated management system. It is essential that environmental issues, with impact on the business opportunities of the company, are treated at a strategic level. In this way improvement activities will be initiated and controlled by top management. Most companies find themselves in a dynamic environment which demands continuous assessment of essential external factors and, where necessary, continuous adjustments to these factors, i.e. proactive planning.

To keep continuous control of profitability, quality and environmental performance it is imperative for an industrial company to have a total management system that addresses issues of a strategic and operational nature. This system must be widely understood, accepted and used by employees at all relevant levels. The efficient use of management systems should complement—and under certain conditions may even substitute for—process and end-of-pipe investments to reduce environmental impacts.

No investments to reduce environmental impacts should be decided before other, more easily attainable solutions have been assessed. It is especially important to prioritize prevention techniques, for example by changing product design/specification or substituting raw materials, wherever such techniques are applicable.

To prioritize company efforts in reducing environmental impacts from products, a life cycle assessment (LCA) approach should be chosen. In this way, both the location and type of environmental impact may be visualized for use in product development and planning processes for environmental improvement.

REFERENCES

Amundsen, A. (1993) *Environmental Technology and Cleaner Production*, Oslo, Universitetsforlaget.
Asbjørnsen, O.A. (1992) *Systems Engineering Principles and Practices*, Maryland, Skarpodd.
Beer, M. and Walton, E. (1990) Developing the Competitive Organisation: Interventions and Strategies, American Psychologist, Vol. 45, No. 2.
Blanchard, B.S. and Fabrycky, W.J. (1989), Systems Engineering and Analysis, Prentice Hall, UK.
Clutterbuck, D. (1991) *The Strategic Dimension*, Henley-on-Thames, Henley Distance Learning.
Fleming, D. (1993) "The Fifth EC Environmental Programme," *European Environment*, special supplement.
Hanssen, O.J. (1994) *Sustainable Product Development—A Draft Method Descriptio*, Fredrikstad, Oestfold Research Foundation.
Johnson, G. and Scholes, K. (1993) Exploring Corporate Strategy, Prentice Hall International, UK.
Senge, P.M. (1992) *The Fifth Discipline*, New York, Doubleday.
SFT—the Norwegian Inspectorate for Pollution Control (1993) *Programme for Cleaner Technology—Evaluation of Environmental Assessments and Demonstration Projects*, Oslo, SFT.
Thompson, F.L. (1993) Strategic Management, Chapman and Hall, London.
Thoresen, J. (1991) *Greenhouse Gas Emissions to the Atmosphere: Causes, Effects and Potential Means of Control of Emissions from Oil and Gas Platforms on the Norwegian Continental Shelf*, University of Stirling, Scotland.
Thoresen, J. (1994) *Environmental Issues Integrated into Management Systems in Industrial Companies*, Fredrikstad, Oestfold Research Foundation.

22
The integrated approach: the Chemical Industries Association's Responsible Care

DR STUART AARON

RESPONSIBLE CARE AIMS

The Chemical Industries Association (CIA) Responsible Care programme was launched in 1989. Its objectives remain unchanged: these are to improve the performance and reputation of the chemical industry in areas that have an impact on people and the environment and thereby secure the industry's licence to operate safely, profitably and with due care for the interests of future generations. The title was taken from the Canadian Chemical Producers Association, along with their support in so doing, because our objectives were the same as theirs.

The CIA Responsible Care programme was based on the guidance generated by the CIA technical and advisory committees from the 1970s. This guidance brought together the best standards and practices of the industry. It also set out the Responsible Care guiding principles to be signed by the chief executive officer (CEO) establishing a commitment to a policy of improvement in all segments to the best industry standard. These principles have been subsequently adopted in similar form by all the 39 countries now signed on to the Responsible Care programme.

In the UK, questionnaires were developed to evaluate whether appropriate health, safety and environmental (HS&E) management was in place. These self-assessment questionnaires examine the clarity of a company's HS&E policy and its

---

Environmental Management Systems and Cleaner Production, edited by R. Hillary.
© 1997 John Wiley & Sons Ltd.

communication, and address the reality of actual operational practice by looking from both the CEO's and the site manager's viewpoint.

One of the key prerequisites of a programme promising continual improvement is measurement. Measurement in HS&E disciplines is largely new since it has traditionally comprised only accident and incident statistics and environmental consent data. The CIA has developed *Indicators of Performance* to monitor the progress in delivering continual improvement in HS&E performance. The *Indicators* have been published annually since 1993, with a growing participation by member companies. They now cover spending on environment protection, safety performance, occupational health, environmental discharges, distribution incidents, energy consumption, product stewardship and communication on these issues.

The commitment to publish the *Indicators of Performance* goes well beyond the scope of existing standards or regulations. It enables the industry to bring real data for the sector to the debate on public concerns. To help achieve this goal, a panel of *Opinion Formers* has been established comprising individuals who are regulators, representatives of pressure groups, financial stakeholders and academics with interests in health, ethics and the environment. The *Opinion Formers* act as a sounding board to advise the CIA on the content and presentation of HS&E issues to the general public and ways in which the voluntary initiative can be moved forward.

Fundamental to the delivery of both credible HS&E performance data and the improvement of that performance are well-designed and effective management systems. The industry will only be able to deliver its commitment to continual improvement in health, safety and environmental protection improvement by the implementation of management systems. Holding the performance improvement gains is one thing, delivering continual improvement is another. Furthermore, a management system is a route to verification or third-party certification, a priority when dealing with many different stakeholders since it gives them more confidence in the chemical industry's commitment.

INTEGRATABLE STANDARDS FOR HEALTH, SAFETY AND THE ENVIRONMENT

The chemical industry is extremely diverse, being comprised of a large range of organizations supplying products and services. Employee numbers span from fewer than 10 employees through to large multinational corporations with tens of thousands of employees. The chemical industry supplies its products and services to most other industries, government institutions and direct to society at large.

This diversity requires the industry to adopt an extremely flexible approach to the demands of its many customers. The industry is also subject to a large number of regulatory controls and industrial codes. The challenge for chemical companies, both large and small, is to obtain, understand and deploy legislation, standards and best practice in a way that is effective for their individual organizations. Companies have, in the past, used a variety of systems for managing and improving business processes. Separate initiatives, including the implementation of new management systems standards, have been taken up in organizations depending on the various business constraints, including quality, health and safety, environment, finance, information technology etc. This piecemeal approach is recognized not to be cost-effective.

It is the industry's view that management systems should be as integrated as is practicable. Indeed, many organizations now view themselves as having only one management system with its core elements satisfying the needs of several management control systems.

This is shown in Figure 22.1 with the six specific requirements clustered round the central common elements of the management process.

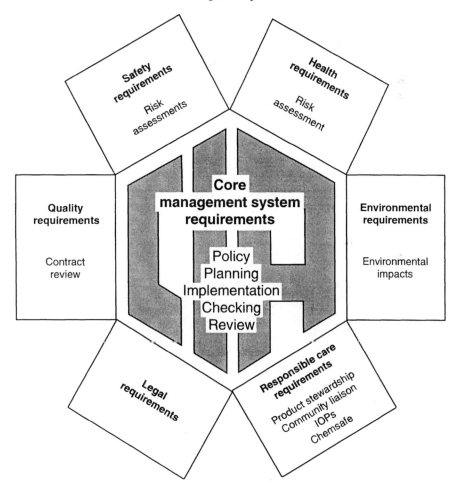

Figure 22.1 Integratable systems

The purpose of such a management system is to ensure that the effects of the activities, products and services of an organization conform to its policy and associated objectives and targets. Furthermore, an integrated system recognizes the fact that very often in industry, and particularly in the chemical industry, the functions of HS&E management are brought together under one individual. Integration brings about much clearer responsibility and involvement.

SECTOR GUIDANCE

In 1992, the CIA pioneered guidance for health, safety and environmental management systems, based on the quality standard ISO 9001, and then took part in various pilots of both this guidance and the then emerging environmental standard BS 7750. The CIA demonstrated together with the major certification bodies, via the series of pilot studies, that in principle this guidance could deliver a certifiable management system and an accredited environmental certification. It was also shown that an integrated system significantly reduces paperwork.

The CIA has recently upgraded this interpretive guidance for an integrated HS&E management system and published it as *Responsible Care Management Systems for Health, Safety and Environment*. Its purpose is to help implement a system such that the activities, products and services of an organization conform to its HS&E policy and associated objectives and targets. The guidance distinguishes the core elements, common to all management systems, from what is being managed.

The Responsible Care guidance has been written in such a way as to address the requirements of:

- The ISO 9000 series of standards (1994);
- BS 7750 (1994);
- Eco-management and Audit Scheme (EMAS) (1993);
- ISO 14001 environmental management systems;
- The Health and Safety Executive's publication, *Successful Health and Safety Management* HS(G) 65 (1991).

Commitment from the top of the organization, the CEO's commitment to Responsible Care, figures at the apex of this scheme, which then drives improvement in the simplified management cycle based loosely on Deming's "plan–do–act–check" cycle. A schematic of this concept is shown in Figure 22.2.

The key elements of this integrated management system include:

- Set objectives.
- Define organization and resource management.
- Identify customer and stakeholder requirements.
- Set targets, prioritize and plan.
- Implement.

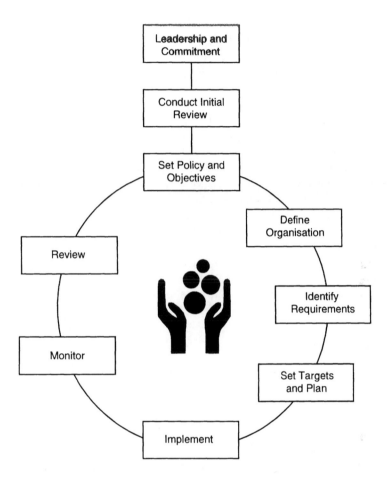

Figure 22.2 Responsible Care continual improvement cycle

- Monitor performance.
- Review.

This approach allows all business functions to perform effectively together in the pursuit of the organization's business goals. In the case of health and safety, for example, the health and safety requirements will be defined by the outcome of the risk assessment. Almost every health and safety assessment will have environmental consequences and vice versa. For example, COSHH and HAZOP investigations and maintenance planning will all generate environmental as well as safety implications. The environmental requirements will be set by the review of environmental effects and the quality requirements will be determined by the process of contract review. Once these requirements have been identified, a common management system may be used to control them all.

The assessment of the significance of the risk should include the entire product life cycle—from product development through to product disposal. Product stewardship seeks to address HS&E risks throughout all stages of a product's life and concerns all parties who design, develop, manufacture, handle or use chemical products down the supply chain. It is an effective way of bringing together such key aspects as supplier performance, manufacturing, waste management, emergency response, warehousing, storage and transportation. The management system needs to identify, document and evaluate as far as possible the actual and potential risks, mainly through hazard identification and exposure elimination. This should be done at each stage of the product life cycle.

The guidance also provides a powerful basis for a self-assessment of Responsible Care implementation since it is based on standards, and this will ultimately pave the way, where appropriate, for independent verification.

ONE-STOP AUDITING

In the same way that an integrated management system can create single work instructions covering all aspects of HS&E, it also makes attainable the concept of one-stop auditing. The industry would appreciate a reduction in the overall costs of certification, valuing, as it does, the real benefits that certification can and does bring. Cost-effective certification could be achieved by multidisciplinary audits which should not be significantly more expensive than a single-discipline visit. This also reduces the costs of a company's own resource input to the audit process, which are far from insignificant.

FUTURE DEMANDS ON THE STANDARD MAKERS

- The chemical industry is seeking an integratable management system standard for occupational health and safety, as argued above.
- New standards must be easily integratable into existing company systems and, as far as possible, new standards should be harmonized with existing ones. This also applies to revisions of existing standards, which should be designed to emphasize the commonality of the core elements.
- Management system standards should address the total affairs of an organization but should be restricted to the minimum number required to achieve this.

- In keeping with the concept of a one-stop audit, the existing audit standards should be harmonized to assist integration of the auditing process.
- Management system standards should include continuous improvement whatever the scope of the standard.

CIA's ultimate goal is a set of harmonized, or integratable, management system standards, internationally recognized, applicable to all aspects of business and delivering accredited certification in a single assessment visit. Certifying the management systems effectively underpins the Responsible Care commitment and thus enhances the public's confidence in chemical industry compliance.

CONCLUSION

The Responsible Care programme, which has now been adopted by 39 countries, has already driven a significant improvement in the UK chemical industry's health, safety and environmental performance and its communication activities. The road forward will be considerably enhanced by utilizing the power of management systems over the full range of management responsibilities. The Chemical Industries Association strongly advocates the concept of integrated management systems and seeks allies in attaining the goal of international, harmonized management system standards.

23
Risk-management approach to environmental management—UK case studies

DR PAUL PRITCHARD

INTRODUCTION

The advantages offered by the implementation of a risk-based environmental management system over a legal compliance attitude can be demonstrated by reference to examples of good practice within industry. Furthermore, a risk-based system can yield additional benefits as follows:

- It utilizes an approach consistent with business thinking and as such can be readily integrated as a mainstream issue.
- It reflects the nature of environmental impacts, where there may be uncertainty as to the nature, scale and likelihood of a specific issue.

There are many common elements shared by an environmental management system, such as those described by formalized environmental management system standards like the international standard ISO 14001, and an environmental risk-management programme. All seek to identify the environmental risks facing an organization, assess their significance and put in place suitable measures for their control. There are, however, some significant differences: for example, risk

Environmental Management Systems and Cleaner Production, edited by R. Hillary.
© 1997 John Wiley & Sons Ltd.

management is likely explicitly to consider risk avoidance (i.e. declining involvement in certain areas because of the magnitude or uncertainty of certain risks) or risk transfer (normally through insurance).

ENVIRONMENTAL RISK-BASED MANAGEMENT

One of the advantages associated with an environmental risk-based management programme is that it closely follows established business thinking. In this way, it can be ensured that environmental issues are given equal prominence to, and more easily integrated with, other business issues. In addition, it can reflect uncertainties associated with the assessment and significance of environmental effects in a consistent and meaningful way. An important recent development is therefore the extension of the remit of risk-management functions within organizations such as Royal & Sun Alliance Insurance Group plc to address environmental issues explicitly. The recent work of the Centre for the Study of Financial Innovation (CSFI) in developing an environmental risk-rating methodology for use by financial analysts is also of interest. This work has been followed by a range of organizations, such as SERM, SYBERR and ECCO CHECK, who endeavour to offer an environmental rating service on a commercial basis.

Eastern Group plc is an example of how a company can reap the benefits of applying an environmental risk-rating system. One of the principal business activities of the company involves the supply and distribution of electricity. Its environmental management system is required to deal with the large number of sites (some of which are unmanned) operated and maintained by the company. In order that environmental staff resources can be focused effectively, an environmental risk-rating exercise was conducted on the sites. This work helps to determine the schedule for the internal audit programme, with sites in more sensitive local settings being given priority. It is also helpful in allocating year-on-year expenditure on equipment upgrades and performance-improvement requirements, which would form part of routine management objectives.

A key element of a risk-management programme is focusing resources where maximum benefit will be obtained, which is illustrated by European Marine Contractors Ltd (EMC), one of the world's leading off-shore pipe-laying companies. When dealing with environmental issues it is subject to national and international legislation that will often be unique to each project. In many cases, a formal environmental impact assessment is required before project initiation. Notwithstanding these requirements, EMC has found it useful to adopt a system within its environmental management programme whereby project managers undertake an early-stage environmental screening. Project managers and other senior staff receive training on environmental issues, which then enables them to apply a screening to their own projects which would seek to identify particular environmental risks and opportunities. This process ensures greater employee involvement and helps to identify those occasions where the input of environmental specialists is appropriate. Focusing resources on an early-stage screening is an effective management tool, since the earlier the issues are identified the easier it is to take appropriate action.

RISK TRANSFER AND THE INSURANCE INDUSTRY

In terms of risk transfer, it is perhaps unsurprising that the insurance industry has been very wary of environmental cover because of the associated difficulties. Experience has shown that it can be extremely difficult to identify when contamination occurred and who was responsible. In addition, clean-up costs are extremely difficult to estimate with a reasonable degree of accuracy and can be extremely high. Gradual pollution presents particular problems in trying to fix the time when pollution occurred and also in providing the connection between the cause (i.e. the gradual pollution) and the effect (e.g. impaired health of claimants). Furthermore, in areas where a range of industries have been operating over a substantial period of time, identifying who has been responsible may not be simple; for example, pollutants can diffuse through groundwater making it difficult to identify a source. It follows, then, that significant liabilities may arise from contaminated land or other adverse events, but that it is very difficult to assess accurately the associated financial implications. In this situation it is very difficult for insurers to offer cover. Given the uncertainties in estimating the (potentially very large) liabilities and the difficulty in transferring risk by means of insurance, it can be seen that internal risk control and reduction become very desirable. Practising risk control is also likely to facilitate insurance cover and help the development of the overall market, since it becomes easier for the underwriter to set a premium. The possible removal of sudden and accidental reinsurance cover from general liability policies in the UK is likely to focus attention on the issue of environmental insurance and could lead to the further development of specialist products.

An example of how a risk-management approach would address both internal control requirements and uncertain environmental effects can be given by reference to a multinational oil company operating in the UK. A risk-based approach can be helpful in the UK because of two factors:

- There is considerable pressure to reuse land which is contaminated.
- There still exist considerable uncertainties in the regulatory framework in relation to the significance of contamination and to what extent it should be remediated.

The risk-management approach involves initial hazard ranking of sites (largely a desk exercise) which then identifies priorities for detailed investigation for soil and groundwater contamination. A site-specific risk assessment can help in developing defensible remediation standards. The UK government strategy, as reflected in the Environment Act 1995, has acknowledged this approach and the more frequent use of risk assessment in this area can be anticipated.

RISK-BASED TOOLS

The British Standard BS 7750 made reference to risk assessment as a possible tool for comparison of identified environmental effects. A suitable approach could involve a matrix-based decision tool whereby the probability and consequence of adverse effects are subjectively assessed and marked on a matrix. The more probable an event and the more serious its adverse outcome, the more the effect increases in significance. An input/output approach can also be applied to identification of environmental effects associated with an organization. Identifying energy and resource flows through an organization is a useful exercise. Risk-based thinking can be consistent with this approach (i.e. routine outputs have a probability of occurrence of 1) and give insight into abnormal and emergency situations where probabilities are lower. An inventory-type approach could make considerations of location and site sensitivity more difficult (unless it were an inventory of risks accounting for location-dependent variation in the consequences of adverse events). Both ISO 14001 and the Eco-management and Audit Scheme (EMAS) are somewhat less prescriptive than BS 7750 in their consideration of risk management. Nonetheless, it is clear that risk-based tools would be of similar utility for both these standards.

A risk-based approach finds no difficulty in dealing with emissions of potentially hazardous substances which are legally sanctioned, for example permitted in a discharge consent. A risk-management approach would not simply accept compliance as a satisfactory position, but would rather view the emission as an ongoing risk which could potentially be reduced. This is a more realistic stance in many ways since it is quite possible that legal standards will change and require action sometime in the future.

As manufacturing chemists Boots Company plc operates in a highly regulated area, but notwithstanding the legal requirements it has found that systematic application of environmental management principles over and above legal compliance yields environmental and commercial advantages. It helps to identify opportunities away from the normal central business activity; for example, the ground maintenance department faced increasing difficulties (both environmental and financial) of storing and disposing of waste generated from the upkeep of the numerous trees and shrubs on the company's premises. A review of the overall position resulted in the purchase of a chipping machine which reduced the timber off-cuts to wood chips *in situ*, vastly reduced the transportation requirements and generated a product which could be subsequently and beneficially used on rose and flower beds to suppress weed growth (and therefore reduce ongoing maintenance). Surplus chippings can also be sold commercially.

ENVIRONMENTAL AUDITS

Environmental audits (within an overall management system) are used to assess the performance of a facility or operation. Organizations often wish to assign values or scores to the audit, thus simplifying the comparison of results across sites and facilitating target setting for performance improvement. Most current scoring systems relate to the implementation of control and procedural measures. An audit methodology could, however, adopt a risk-based approach, which would consider the operation as having three elements related to environmental impact:

1. Hazards, which may be chemicals or processes.
2. Controls, which may be physical or managerial.
3. Receptors, which are the targets of emissions (local or global).

This approach emphasizes the view that management control elements play an important role in determining the overall level of risks, but these must be viewed in conjunction with less controllable components, namely the inherent hazards and the potential impacts on receptors. In this approach the inherent hazards are subject to controls which minimize the risk to receptors. Physical control elements, such as pollution-prevention devices, are clearly also subject to site management control.

CALCULATION OF RISK PROFILES

A clear management strategy to minimize the total risk could then result from the calculation of individual site risk profiles. This should direct finite resources to where they can be most effectively deployed and could in itself be used as a key corporate performance indicator. At a basic level, simple high, medium or low (H, M, L) assignments could be made for both the inherent hazard and the receptor categories. Assigning scores of 1, 2 or 3 to low, medium or high, the overall site risk can be calculated by combining the three elements:

Hazard score × control score × receptor score = risk profile

Application of this scheme would enable organizations to retain existing audit scoring mechanisms, since these methodologies largely focus on management and physical control elements. Higher hazard and receptor scores would then need to be counterbalanced by higher control scores.

The hazard assignment (H, M, L) would depend on the assessment of materials used and the processes conducted on the site based on criteria already identified. The control element could be the score from an existing audit system, the receptor score would then be based on site-specific characteristics such as nearby population, geology and hydrogeology.

Risk assessment was also described in BS 7750 in the context of evaluating environmental effects arising out of emergency conditions, where the probability of an adverse event occurring and the severity of the environmental effects arising out of the incident both need to be considered. This assessment would then form the basis for establishing objectives, controlling risks and establishing environmental components of emergency plans. This approach can be seen as extending the traditional view in health and safety studies by considering a broader range of materials to be hazardous (e.g. stratospheric ozone-depleting substances such as CFCs) and considering eco-system targets in addition to humans.

Rockwater is a major off-shore support company that has supplemented its considerations in this way. Furthermore, it also integrates environmental issues within other health and safety requirements such as the self-assessed risk of routine operations. Although a minor spill of several litres of oil, for example, could in no way be considered a major incident, it is inherently more likely to occur. A simple subjective risk assessment (possibly by the operative in person) offers the potential to reduce overall environmental effects by minimizing the probability and consequences of such relatively minor events.

CONCLUSION

The application of formal environmental management systems is extremely useful in achieving improved environmental performance. In many cases, however, the improvements gained may be subjectively assessed or have significant associated uncertainties. It is therefore appropriate to end with an example demonstrating quantitative improvements gained from environmental management system implementation.

Courtaulds plc's worldwide businesses are grouped into three areas: coatings and sealants, polymer products and fibres and chemicals. Each shares a common route in the technology of polymers. Most Courtaulds UK businesses are to quality standards in the ISO 9000 series registered and have developed environmental management procedures; the appearance of BS 7750 was considered timely and could form the basis of a consistent approach throughout the company.

Within Courtaulds, Amtico led the way in the development of a formalized management system. Amtico is part of the polymer products group and manufactures luxury floor tiles for houses, stores, offices and hotels. Manufacture takes place in Coventry, UK and is supported by a worldwide network of design studios.

Amtico created an environmental advisory panel, the members of which included the senior director responsible for environmental matters and a "green group" representative, an employee chosen for a known interest in environmental issues. The panel, building on the business's ISO 9000 experience, has led Amtico through the steps of environmental policy, organization, register of effects, objectives, targets and a programme, documentation, training, controls, records, auditing and review.

Amtico has already derived benefits from the application of the systematic approach to environmental management. In the year following implementation it saw energy costs down 6%, water bills down 13% and waste disposal charges reduced by 7%.

REFERENCE

Pritchard P. (1994) *Managing Environmental Risks and Liabilities*, Business and the Environment Practitioners Series, Cheltenham, Stanley Thornes.

24
Experience of environmental management in the Danish fish-processing industry

ESKILD HOLM NIELSEN

CLEANER PRODUCTION IN THE FISH-PROCESSING INDUSTRY

In 1987, the Cleaner Production Programme was established in Denmark. The purpose of the programme was to develop cleaner production projects and disseminate them in Danish enterprises. This enabled the companies and their networks to receive economic support in order to implement development and demonstration projects and, to a certain extent, information activities. The programme only provides support for selected industries including the fish-processing industry. In Denmark the concept of cleaner production is used in a restrictive manner which excludes "add-on technology" and (external) recycling activities.

In the period from 1988–92, the programme spent Dkr 24 million on cleaner production projects in the fish-processing industry. Of this sum, Dkr 6 million went on expository and informative projects. The remaining Dkr18 million was used on cleaner production projects, primarily within the herring industry. The following cleaner production projects have been completed in the herring industry under the Cleaner Production Programme:

Environmental Management Systems and Cleaner Production, edited by R. Hillary.
© 1997 John Wiley & Sons Ltd.

1. Consultancy schemes for the fish-processing industry in North Jutland.
2. Herring gutting by scraping/suction.
3. Cleaner production at all stages of the production process.
4. Reuse of basic and acidiferous solutions and improved handling of herring barrels.
5. Use of enzymes from herring guts in the curing of salted and spiced fillets.
6. Improved utilization of process water in the herring-filleting industry, by means of purification, disinfection and recycling of the process water.

With the exception of the consultancy schemes, the projects were implemented in the same company as demonstration projects. Efforts in the herring industry have primarily been concentrated on the development of cleaner processes, with the aim of solving specific environmental problems related to the industry.

In the consultancy schemes, Project 1 had the general aim of expanding the knowledge of cleaner production and assessing the possibility in the individual company of adapting cleaner production. This project is the only one which focuses on the general environmental performance of the industry. The project should be regarded as one out of many initiatives aimed at improving the dissemination of cleaner technologies. The dissemination of experience in cleaner technologies obtained through the programme has shown itself to be more difficult than was first realized (Christensen 1993 and Andersen 1994).

Project 2 had two parts, each dealing with the dry transport of herring guts. The purpose of the project was to avoid contact between process water and the guts. By use of dry transport, a larger proportion of the guts becomes marketable waste and pollution is reduced by preventing any part of the gut mass being dissolved in the process water. The project revealed that there was an environmental and economic advantage in scraping the guts away instead of using suction. This technological optimization of the filleting process has since been implemented in many companies in the herring industry.

Project 3, "Cleaner production at all stages of the production process", was thought of as a consolidation and further development of cleaner initiatives in the industry. The project involved the reduction of water consumption at all stages of the production process, collection of waste in a dry state, mechanical herring transport instead of water borne, removal of skin from the skinning machine by suction, impulsing water on the machines and improved cleaning. This project, along with the automatic handling of the barrels, was continued in full-scale implementation.

The remaining cleaner production projects were not suitable to be carried out in practice. Project 6 had not been completed when the assessment was carried out and is not included. The projects are very technologically oriented, i.e. the solving of the environmental problems is directed towards a "technological fix".

DESIGN OF THE CASE STUDY

The aim of the study was to determine to what extent implementation of the cleaner production projects has resulted in an actual reduction of pollution from the herring industry. The part of the examination included here has two aspects. First, a qualitative assessment of the experience gained from the many cleaner production projects which were carried out. Second, an examination of the development of environmental pollution from five herring-filleting enterprises in the local authority district of Skagen, North Jutland, in the period 1989-94, with the aim of assessing the environmental effect of the cleaner production initiatives in the industry at large. These enterprises have been responsible for the greater part of the organic matter in the district's waste water.

The degree of environmental pollution is assessed on the basis of changes in the companies' water consumption and the discharge of organic matter (measured as kg COD (chemical oxygen demand) per ton of raw materials). The COD discharge expresses the pollution from the companies and the water consumption and waste volume express their resource awareness. The changes in the two parameters are related to the implementation of cleaner production projects. The two parameters are reckoned per ton of raw materials, for the purpose of eliminating the importance of variations in the volume of production.

Much of the data which should have been used in the assessment was not readily available from the companies, but they provided us with permission to obtain the information from various public authorities and other sources. The Water Authority supplied information about the volume of water consumption and the fishmeal factory informed us of the amount of waste. The local authority provided results from waste water tests, which also contained information on COD concentration. The only quantitative data provided by the companies was their consumption of raw materials. Because the data had to be related to the actual technology, each company was visited and inspected and interviews were carried out with those responsible for environmental matters.

THE ENVIRONMENTAL EFFECT OF CLEANER TECHNOLOGIES IN THE HERRING INDUSTRY

The following section describes the improved environmental performance of the companies and relates it to the expected effect of the cleaner production projects which are suitable for implementation. Table 24.1 indicates the consultants' expected reduction of pollution from these projects. The reduction achieved for cleaner production at all process stages is based on development projects, not on the basis of full-scale implementation.

Table 24.1 The expected effect of the cleaner production projects

Cleaner production initiatives	Cleaner production in all process stages	Scraping
Reduction in water consumption	50%	0%
Reduction in COD discharge	60%	30–40%

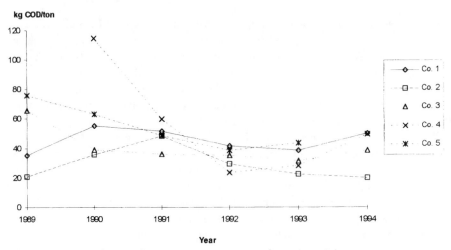

Figure 24.1 Discharge of organic matter per ton of raw materials

Figure 24.1 shows that there was a considerable variation between the companies in their degree of pollution in the period 1989–94. This variation in COD discharge was reduced until the period 1992–94, so that the herring processing companies today have a much lower and more similar discharge. The variations in COD discharges had a factor of 4 before the dissemination of the implementation of cleaner production projects, but this was reduced to a factor of 2 in 1994.

The falling COD values from 1992–94 in company 2 were a result of cleaner production in all stages of the production process. For company 5 this project also reduced the COD values from 1993–94. Water consumption fell from 1992–94 at a faster rate than the rise in COD concentrations. Therefore the total COD discharge from both companies fell.

The cleaner production initiatives have resulted in a greater knowledge of cleaner production and the industry has taken a technological direction, e.g. three out of five companies have introduced gutting by scraping and all five have introduced filter belts. After completion of the study the remaining two companies also adopted the scraping method.

"Cleaner production at all stages of the production process" (project 3) was completed at the end of 1992 and one further company (in addition to the project-implementing one) has carried out a similar project. Almost all the companies, to a certain degree, have implemented elements of this project.

The remaining variations in the companies' COD values can primarily be explained by an inadequate dissemination of project 3 and differences in the companies' application of abatement technologies, e.g. centrifuges, flotation. Furthermore, the companies practise different types of production planning, and there have been differences between each company in the degree of action taken to reduce the pollution which resulted from working procedures. This final aspect is discussed in more detail below.

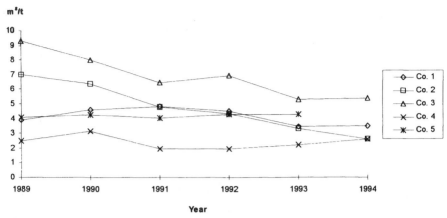

Figure 24.2 Water consumption in the herring industry

Figure 24.2 shows that the water consumption per ton of raw materials in the herring industry has decreased, but there is great variation in the companies' starting points and development of water consumption. The reduction in water consumption can be explained by two reasons. First, the companies have all implemented water-saving devices (better housekeeping). Second, all five companies have increased production, which enables the economies of scale to be effective because the basic water consumption is constant. The basic water consumption is that part of the volume which is independent of the size of production, e.g. cleaning, tub filling for the untreated herring. During the period there was also an increase in the degree of product refinement, in the form of descaling/skinning and more products being sold in brine or pickled, so there has been an increase in water consumption.

Despite this, several companies succeeded in reducing their water consumption during the period 1989–94. By means of better housekeeping, company 3 reduced its water consumption from over 9 m^3 to approx. 5 m^3, and the increase from 1991–92 was the result of the introduction of skinning. The higher the starting point, the easier it was to reduce the consumption. Company 5 held a constant, relatively low, water consumption until 1993. The implementation of various water saving devices halved their water consumption from 1993–94. The results from 1989–91 of company 2 can primarily be explained by a rapid increase in production. The remaining decreases for companies 1 and 2 can primarily be explained by cleaner technologies, where the companies implemented most of the changes from the experiences gained from project 3 ("cleaner production at all stages of the production process"). The reduction from 1993–94 for company 2 was related to its work to establish an environmental management system. Reduction of water consumption was possible in the two companies with a higher degree of product refinement. Company 4 has a very low water consumption because of its lower degree of product refinement.

Generally, it must be stated that the companies have improved their environmental performance. Collectively, they have almost halved their COD discharge, and reduced their water consumption with about 40% per ton of raw materials. These improvements are primarily due to the implementation of various cleaner technologies. The cleaner production projects are the results of the Cleaner Production Programme's initiatives in the field. However, the companies have not satisfied the expectations indicated in Table 24.1. This can be explained by an increase in product refinement and the technological optimism of the consultants. The development and dissemination of the various cleaner production projects have contributed to the improved environmental performance of the companies and reduced the variations between them. A more detailed discussion of the results will be given below.

DISCUSSION

The companies wented to reduce their water consumption and environmental pollution because of the more severe controls to which they are exposed. Implementation of the cleaner production projects has not been dependent on economic incentives. There are two reasons for this. The companies have become environmentally more aware and they wish to be prepared for the future tightening of environmental demands on the industry (Christensen 1993). Primarily, the implementation of cleaner production projects in the companies must be recognized as anticipating future environmental regulation.

In the companies studied, there has been a desire to reduce the environmental loads, both in relation to water consumption and organic matter. Their environmental strategies have been technologically oriented because all the implemented devices have been technical. Only company 2 has included an environmental management system in its environmental strategy. The environmental improvements in the form of reduced COD discharges can primarily be related to technological improvements and secondly to improved cleaning procedures.

Technologically, the companies are homogeneous and at the same time their products are identical, therefore it would be expected that their water consumption and COD discharges would be the same. This is not the case as there is still a large variation. As an illustration, the company that has implemented the cleaner production projects can be regarded as the most modern, from a technological point of view. Even if this company is in the vanguard technologically, its environmental performance is mediocre. It has one of the largest COD discharges and has only average water consumption.

The technology in use in this company must be regarded as the best available within the industry. Its environmental performance is mediocre, therefore it is necessary to differentiate the sources of pollution into two types. Pollution can be related first to the applied technology and second to the layout of the company, as well as to the working procedures of the employees and to production planning. The Danish cleaner production strategies, and the strategies of the fish-processing industry, have both focused on that part of pollution which can be related to technology.

Technologically, the company in which all of the cleaner production projects were implemented has solved its environmental problems, but it has failed to concentrate on the pollution which is related to its working procedures and production planning. The only thing that it has done in this connection is to encourage the cleaning contractor to use less water. By requiring the cleaning contractor to register its daily water consumption and to deliver this information to the production manager, the water consumed was reduced from 120 m^3 to 80 m^3. Internal recycling of cleaning water could have reduced the consumption by a further 30 m^3, but this was not done until after the study. In this particular company, it was recognized that the total water consumption for production can be reduced from 3.3 m^3/ton to 2.5 m^3/ton, with the aid of better production planning and involvement of the employees. In Denmark, there are many examples of employee involvement resulting in significant environmental improvements (Nielsen and Remmen 1994).

The cleaner production initiatives in the fish-processing industry must be regarded as a success, from a technological point of view, because many of the projects have been implemented in the demonstration companies, and because the initiative has had a secondary environmental effect in the rest of the industry. It should be stated that the dissemination of cleaner production projects has been higher in Skagen, which probably can be explained by better knowledge of cleaner technologies and better implementation of environmental regulation (Christensen 1994 and Andersen 1994). The cleaner production strategies of the companies and the National Environmental Protection Agency must be supplemented with action directed towards that part of pollution which is related to working procedures and production planning. This can be effected by means of environmental management systems and more "dynamic" regulation, directed against all sources of pollution.

Of the five companies, only company 2 has included an environmental management system. 1992 was the starting point for environmental management and the company was certified in May 1995. The company wanted an environmental management system in order to get a better relationship with the regulatory authority and to be prepared for future forecast tightening of environmental demands (Jørgensen 1995). From 1992, the company worked extensively on its environmental problems, especially water consumption and discharge of organic matter. It managed to reduce its discharge of organic matter (kg COD/ton raw material) from 48 kg to 20 kg in 1994. The company expects to reduce the discharge even further to 16 kg in 1996 (Jørgensen 1995). Correspondingly, company 2 was able to reduce its water consumption from 6.3 m^3 in 1990 to 2.6 m^3 in 1994, and in its action plan it states that in 1996 water consumption will be less than 2 m^3 (Jørgensen 1995). The company has also included other environmental aspects in the system.

When the environmental management systems were being constructed the employees were more eager to implement environmental improvements than to build up the system (Jørgensen 1995). The improvements were oriented towards the entire pollution load. The employees' eagerness and the company's environmental strategy led to environmental improvements during implementation of the environmental management system beyond what is stipulated in BS 7750. For example, BS 7750 does not require implementation of environmental improvements from the start of the EMS. But the company was very anxious to improve its environmental performance right from the beginning of the environmental management process. In May 1995, 15 companies were certified to BS 7750 (Christensen and Nielsen 1996).

It must be stressed that company 2, with the environmental management system, had the best environmental performance among the five companies in 1994 and it seems likely that they will preserve this position. Although company 2 had a mediocre environmental initial position in 1990, it has been able to change more easily than the other companies in Skagen. This position is first of all linked to the fact it has implemented an environmental management system which has resulted in many environmental improvements. The establishment of the environmental management system has created a more proactive environmental strategy that goes beyond the traditional environmental regulations and the cleaner production strategy of the other companies. The proactive strategy is both offensive and holistic towards environmental problems.

If future cleaner production strategies are to have a broader environmental effect, they will have to include environmental management systems to a greater extent, in order to ensure that prevention of pollution includes the pollution which is related to working procedures and production planning. The Cleaner Production Programme has provided economic support for the implementation of environmental management systems in four companies, but it is difficult to envisage how such systems can be introduced on a wider scale.

In recent years, the term "environmental benchmarking" has been introduced (Williams 1994). However, there have not been many proposals as to how this should be defined. In connection with the concept of integrated pollution prevention control (IPPC), it is possible to see the contours of how "best available technology" (BAT) can be defined. There are no tools to determine environmental benchmarking for pollution which cannot be related to the applied technology. This and other studies (Nielsen and Remmen 1994) have shown that it is important for environmental benchmarking to be directed towards all sources of pollution. It is not sufficient to rely on technology alone. Environmental benchmarking must also include pollution which can be related to production planning and working procedures. Benchmarking must be related to the actual environmental performance of the industry. Furthermore, it must be related to environmental regulation.

On the basis of this study it must be recommended that strategies for cleaner production be linked to environmental management systems. Environmental management systems can be considered as a key element in disseminating cleaner production projects and better housekeeping. This study also shows that companies implementing environmental management systems have great possibilities of reducing all sources of pollution, including the pollution related to working procedures and production planning.

REFERENCES

Andersen, M.S. (1994) "Spredningen af renere teknologi i den danske fiskeindustri", *Arbejdsrapport*, NR. 41, Miljøstyrelsen.
Christensen, P. (1993) *Spredningen af renere teknologi i den nordjyske fiskeindustri*, Aalborg University.
Christensen, P. (1994) "The influence of the implementation process on the success of the environmental regulation: a case study of the Danish fish processing industry", *Journal of Clean Technology and Environmental Sciences*, Vol. 4, No. 1/2, pp. 79–99.
Christensen, P. and Nielsen, E.H. (1993) "Environmental audits, cleaner technologies and environmental protection in Denmark", *European Environment*, Vol. 3, Part 2. pp. 18–22.
Christensen, P. and Nielsen E.H. (1996) *Implementing Environmental Management Systems in Danish industry—Do we go beyond compliance?*, Eco-Management and Auditing, Vol. 3, No. 2, pp. 56–62.
Jørgensen, T. (1995) *Kvalitets- og miljøstyring på Erik Taabel Fiskeeksport A/S—Integreret kvalitets- og miljøstyring i levnedsmiddelindustrien*, Institut for Samfundsudvikling og Planlægning, Skriftsserien nr.177, Aalborg University.
Nielsen, E.H. and Remmen, A. (1994) *New Incentives for Pollution Prevention*, paper presented at Third International IACT Conference, Vienna.
Williams, Marcia (1994) "Why—and How To—Benchmark for Environmental Excellence", in *Understanding Total Quality Environmental Management*, New York, Executive Enterprises Publications.

25
Cleaner production through environmental management of process innovations

NILS THORSEN

INTRODUCTION

The purpose of this chapter is to address the question of whether the current level of cleaner production and environmental management systems in Denmark and throughout Europe is sufficient to overcome the evident challange for industry in the 1990s to realize ecologically sustainable growth. To answer the question, this chapter focuses on the realization of cleaner production through environmental management systems from the specific view of process innovations.

Further questions arise. Can a strategy for environmental management systems focused on the realization of cleaner production through process innovations be widely implemented throughout industry? Will the strength and comprehensiveness of environmental protection allow us to hope that the challenges of ecological sustainability can be accomplished? The empirical evidence in this chapter is from Danish industry and indicates that the strategy is attainable.

FUNDAMENTAL CHANGES IN THE GENERAL ORGANIZATION OF ENVIRONMENTAL PROTECTION

Industry strategy is to consider environmental management as a natural part of the general company management. This is the beginning of fundamental changes in the

---

Environmental Management Systems and Cleaner Production, edited by R. Hillary.
© 1997 John Wiley & Sons Ltd.

general organization of environmental protection and the principal roles of the actors involved.

The profile of a company's pollution prevention strategy is altering radically. From being reflected images of requirements from environmental authorities, the profile is altering into a profile grown from visions and management strategies. This new approach from industrialists is considered as an adequate response to the messages and the latest legislative initiatives from the central authorities, the Danish Environmental Protection Agenecy (EPA) and the Ministry for the Environment. The message is that the obligation for initiatives and responsibility in environmental matters is expected to move from local authorities to the companies themselves. Examples of such legislative initiatives are the Eco-management and Audit Scheme (EMAS) and Eco-labelling regulations, green taxes and the new Danish law on green accounts (2000 Danish companies are obliged to produce green accounts/statements parallel to the financial accounts from 1997 including 1996 figures).

The local authorities' attitudes to these facts vary from fear to a proactive and constructive response. As regards the proactive response, local authorities are developing new ways of local regulation and inspection, which fit into the progressive business approach. New inspection opportunities are based on audits focused on management initiatives to prevent material pollution at source. These audits successively replace inspections and monitoring of end-of-pipe pollution to the source of pollution. New regulation opportunities are now based on flexible framework approvals implementing continual environmental improvements through a linkage to the company's environmental action programmes where cleaner production opportunities are applied as they emerge. These forward-directed framework approvals successively replace traditional approvals based on detailed end-of-line requirements linked to threshold values for groups of single hazardous substances. Threshold values are often fixed for a decade or more despite the fact that new abatement technologies or substitution opportunities are developing.

The aim of this cooperative strategy for business and authorities is to raise sufficient resources in companies as well as in local authorities to promote real preventive measures. Taxpayers and consumers cannot be expected to pay for these measures on top of the existing budgets for environmental protection. Consequently, it will be necessary to mobilize resources already spent on controls on detailed, narrow limits and expensive end-of-pipe emission-control programmes. These programmes do not often reveal a broad and continuous picture of environmental performance.

THE ACHILLES' HEEL OF ENVIRONMENTAL MANAGEMENT SYSTEMS

The Danish Environmental Protection Act includes an obligation for the companies to utilize cleaner technologies (a term which in this chapter will be used interchangeably with cleaner production). Cleaner technologies should be applied to ongoing process innovations as far as these involve application for new environmental approvals. The fulfilment of this regulatory requirement has been facilitated by a comprehensive cleaner technology programme from the Ministry for the Environment, approximately £10 million sterling per year from 1986. This strategy has been reasonably successful.

From a cleaner technology perspective, environmental management systems are not a guaranteed way of achieving the choice of the best technology. Despite Article 3a of EMAS, which obliges companies to a policy commitment to the "economically viable application of best available technology" (EVABAT), environmental management systems tend to be "without level" regarding the the strength of environmental protection. Without question companies do have to decide which environmental protection level should constitute the basis of their management system as long as regulatory requirements are met. "Continual improvement", the policy obligation in the environmental management system standards ISO 14001 and BS 7750, is not very well defined. The management system guarantees that the chosen level of environmental protection is always targeted, although the system does not guarantee the choice of cleaner technology. To some extent there is a theoretical guarantee of this choice in Denmark because of the existence of the cleaner technology obligation for process innovations.

MEETING GLOBAL ENVIRONMENTAL CHALLENGES

The increase in natural resource consumption and pollution is a burden and threat to global ecological sustainability. Consumption of natural resources and the production of pollution mainly depend on population growth, the standard of living and the technologies used. To create a new sustainable development model the only option is to strive to change the technologies in use. Mandatory reduction in population growth or living standards is not a real consideration. Therefore, cleaner technologies are the option that needs to be realized. With the current growth in population and the increase in living standards much needs to be accomplished in cleaner technologies if the global burden on nature is to be decreased or at a minimum stabilized.

However, experience with cleaner technology tells us that most projects are "one-night stands": they are merely one-off stand alone projects, often related to support grants from cleaner technology programmes. Cleaner production opportunities are not spread to industries in general.

In this respect, management systems offer an opportunity as the system must include plans for the management and control of all environmentally critical processes. Proposals are presented in this chapter for environmental management systems which focus on "best environmental improvement per unit effort". Such systems will naturally focus on the procedures for process and product innovations realizing cleaner production in business. Environmental protection can be improved in "leaps" in these areas.

IS CLEANER PRODUCTION COMPETITIVE?

The $10 000 question is if business has any strong imperative to follow a cleaner production strategy. The existence of such imperatives is—in a liberal economy—of course a precondition for actions along that line to be taken in practice. Fortunately, Danish experience gives us some hope.

Even if initially environmental protection costs, it is reasonable to conclude that Danish businesses have enhanced their competitiveness in parallel with environmental improvements by stimulating market opportunities for environmentally sound products and equipment.

The Danish Ministry for the Environment has attempted to ensure that polluting industrial plants fulfil international environmental requirements early. This strategy serves Danish industry with a "first-mover" advantage with respect to competitiveness, which primarily emerges when competitors in other countries are obliged to meet the same requirements.

Danish industry has worked with an approval system for polluting plants for 22 years and—as mentioned above—in the latest 10 years approvals have been based on environmental standards equivalent to clean technologies or best available technologies (BAT). Based on the BAT evaluation, an approval system is now being developed in the European Union. The EU Commission has, in principle, agreed a directive on "integrated pollution prevention and control" (IPPC). An example of the competitive advantage stimulated by Danish environmental policy is that Denmark is two years ahead with respect to substitution of CFCs. This has been achieved based on a mix of measures, including regulation combined with levies on CFCs to finance grants for developing alternatives. Danish producers of district heating pipes are first on the world market with CFC-free pipe insulation foam and its producers of refrigerator compressors are the first on the market with compressors which can operate without CFC.

Investments in environmental improvements can be equivalent to enhanced competitiveness, for example when waste-reduction measures represent savings on raw materials or purification costs, especially when achieved with a rapid payback period. It is often acknowlegded that environmental requirements alert company management to unutilized resource management opportunities. The National Danish Action Plan for Water Environs has resulted in investments in fishery industries and slaughterhouses that reduce water consumption by up to 80% by simple and cheap means. New emission standards have resulted in the surface treatment industry reducing the consumption and emission of organic solvents from paints by new technologies, organizational changes and internal recycling systems.

EXAMPLES FROM BUSINESS PRACTICE: MANAGING PROCESS INNOVATIONS TOWARDS CLEANER PRODUCTION

The following examples all represent existing practice in Denmark.

Example 1   Opportunities to manage process innovations in a large-scale chemical industry

In a large Danish company in the chemical sector, the business includes different business areas supplying different markets. Quality management systems have penetrated most business areas. The company has its own engineering department which is constituted as an independent business area in the group and concerned with most process innovations company wide. The engineering department has had a quality management system for some years, although the system is not yet certified. The engineers generally consider a number of critical factors connected to process alteration or the implementation of new processes. As production processes are highly automated, the level of environmental protection is very much dependent on equipment and facilities.

The corporate health and safety department has succesfully integrated regulatory requirements concerning safety plans for construction work in the engineering department quality system. Success was obtained by "translating" the substantial elements of a legal Act from the Danish Ministry of Labour concerning the safety plans to the quality system "language", and then adopting the text in the quality manual. As the engineers are generally guided by the quality manual in day-to-day operations, it was natural to include the obligations concerning the safety plan for construction work. This rather simple approach has achieved good results in compliance with the health and safety regulations and increased safety in construction work.

It was suggested that implementing the regulatory requirement of the Danish Environmental Protection Act concerning cleaner technologies would be a natural element of good engineering quality. This would be possible by simply adding it to an existing quality procedure for "analysis and consideration of critical factors for the project". This addition to the quality procedure is being considered. It should be emphasized that additions to the procedure will bring about alterations in several detailed work instructions applied in the engineering departments.

Example 2   Employee participation in process innovations by a medium-sized manufacturer of glue

A medium-sized manufacturer of industrial glue had to deal with health and safety and environmental problems in the processing of glue. Cleaning process tanks between each batch of different items of glue was causing problems.

The cleaning process included manual washing with organic solvents of the process tank and a steering propeller after each product batch. The workers used comprehensive ventilation and safety equipment, but the working procedure was still not completely safe. After the cleaning process, considerable amounts of hazardous waste needed to be collected and disposed of to the Danish Chemical Waste Treatment Plant, KommuneKemi, at a total cost of approximately £500 000 per year.

Several years earlier, the company had taken part in a project with the Organization of Danish Industries concerning employee participation in process innovations and had also received the ISO 9001 certificate for quality management. Throughout these two initatives the organization had developed a tradition of rapid changes to processes which actively involved employees. The Kaizen quality principles and experiences with "quality circles" were applied. The process included group meetings and discussions concerning new business goals and the implementation of proposals to achieve the goals. These meetings, including operators from the shopfloor, were scheduled as part of the normal working plan.

During an environmental review the issue of the cleaning process was investigated. Costs linked to consumption of solvents and waste disposal, health and safety problems and delays to start-ups of new product batches caused by time-consuming cleaning were causing considerable problems. Objectives were set up to solve these problems and a process of change was initiated. After several meetings, many proposals were suggested and evaluated with many being withdrawn, until finally the solution was found. It was simple, came from the shopfloor and was the most efficient solution.

Before feeding the process tank with raw materials a large plastic bag was placed in the tank (at the start it was blown into the tank with a reversed vacuum cleaner). The stirring propeller was taped with solid tape. Processing was then accomplished normally, but after emptying the glue mixture into a storage tank the workers waited 10 minutes until the glue residue had hardened. The plastic bag could be removed and the tape on the steering propeller loosened with a pocket knife once all the glue had hardened, and plastic and tape could be disposed of at an incineration plant.

Annual disposal costs of hazardous wastes declined from £500 000 to around £100 000. Consumption of solvents also declined. Occupational health problems were minimized and new product batches could be started more quickly. The company offered the team behind this process innovation a special grant. The production manager identified one problem with the project, the fact that the the company had been processessing in the old way for so many years. The company was among the first in Denmark to be certified to BS 7750. Manufacturing practices at the company already went beyond the standard requirements, therefore it did not find the implementation of BS 7750 so difficult to achieve.

Example 3 Environmental management of consultancy

For small and medium-sized enterprises (SMEs) in particular, process innovations often depend, to a large extent, on the competence of professional consultants or suppliers of equipment. SMEs often lack competence to undertake major alterations to their processes or facilities. As the Danish industry structure is dominated by SMEs and as pollution often originates from this category of enterprises, environmentally sound services from consultants are essential for general environmental protection.

A major Danish consultancy firm was offering a broad range of industrial services including process engineering and construction, also environmental services including environmental management. This consultancy had decided to implement an environmental management system in its own business. Initially the environmental management system was directed to the consumption of water, paper, energy, space and materials in the process of consultancy itself. Under the guidance of the environmental department, the question of system efficiency was put on the agenda. It soon became obvious that focusing on the consumption of paper, energy, water and space facilitated improvements in principle, but substantially offered a very limited contribution to sustainability. On the other hand, focusing on the enviromental soundness of the company's services offered an opportunity for major environmental improvements and pollution prevention.

Therefore the core of the environmental management system became procedures to ensure that the service products considered environmental quality. This meant that the company's technicians were to carry out evaluations of the environmental consequences of delivered service or advice and investigate possible cleaner and more sustainable alternatives. Good consulting practice was to be always to inform the customer of the intrinsic environmental aspects of the required service products and possible better alternatives. However, it was not up to the consultant to make choices for the customer, as the final result depends on the customer's imperatives and interests. In a wider perspective, it was also up to the customer to decide on willingness to pay for the extra investigations of environmental consequences and opportunities.

CONCLUSIONS

If environmental management systems live up to today's ecological challenges, they must be able to promote radically cleaner production. Cleaner production can be obtained throughout better housekeeping in day-to-day operational controls, but radical improvements are often connected to process innovations. Therefore environmental management systems should focus on the companies' procedures for process innovations. Three Danish cases indicate that this is possible.

The case of a major company in the chemical sector shows that engineers are able to handle the environmental issue as competently and rationally as any other issue if they are adequately managed and guided. The case of the medium-sized manufacturer of glue indicates that an active and inventive organization can realize environmental improvements as well as any other improvements. The example from the consultancy demonstrates that this sector also wishes to launch "green" process innovations.

However, even if these opportunities exist, it is too early to conclude whether they will be widely utilized. This depends on thousands of management decisions. These management decisions in private enterprises principally depend on competitive advantages, profits and market shares. The fundamental question seems to be: do cleaner production and ecological sustainability equate to increases in market share, profits and competitiveness?

Section V
Industrial Experience from Emerging and Transition Economies

26
Introduction

RUTH HILLARY

The increased globalization of trade and harmonization of standards require that enterprises from the north, south, east and west need to consider more effective ways of managing the environmental aspects of their operations. In this, the sixth section of *Environmental Management Systems and Cleaner Production*, the challenges facing industry situated in emerging and transition economies is considered. In its five chapters the section draws on case studies from South Africa, Brazil, Lithuania, the Czech Republic and Hungary and from the oil, mining, steel, energy and water supply, textile, chemical, metal-finishing, plastics, furniture, pharmaceutical, food-processing and construction sectors. Central themes of this section are the interrelationships between the changing political and social fabric of a country and its economic development and the implications of these relationships for enterprises newly exposed to the international market and its standards.

In Chapter 27, Harold Nicholls considers how legislation, environmental management systems, environmental capacity building and awareness campaigns are all integral parts of an environmental management system for cleaner operations in South African mines. He reviews the legislative framework which requires mining operations to seek authorizations and produce an environmental management programme report (EMPR), asserting that a shortcoming of the law is its lack of contribution flexibility to an environmental fund for mine closures. Nicholls draws out the essential ISO 14001 environmental management system components found in an EMPR. He points out the importance of staff support and understanding of their individual obligations, detailing a programme of interactive workshops designed to secure mine employees' involvement. Nicholls concludes that commitment, an environmental management system and staff capacity building are all of equal importance to achieve cleaner operations in mines.

Sergio Pinto Amaral, in Chapter 28, discusses the detailed case study of the Brazilian state-owned oil industry monopoly Petrobras and its environmental

---

Environmental Management Systems and Cleaner Production, edited by R. Hillary.
© 1997 John Wiley & Sons Ltd.

initiatives, seeking to show the links between environmental management systems and the promotion of cleaner products and processes. He discusses international standards and the tendency for Brazilian legislators to make voluntary environmental management tools, such as environmental auditing, mandatory. Furthermore, he argues that the legislators are reacting to the weakness of the official environmental control agencies by trying to transfer the responsibility for environmental control to the production facilities. Amaral provides an insight into the workings of the environmental management system in Petrobras, detailing key environmental initiatives such as the addition of ethyl-alcohol to gasoline, the use of low-sulphur diesel in large urban centres and environmental awareness raising and training. He concludes by suggesting that it is the combination of company awareness of environmental issues and demands from society, consumers and government and non-governmental organizations (NGOs) which is driving environmental issues in the company.

In the first of three chapters which consider the industrial experience from Central and Eastern Europe (CEE), Leonardas Rinkevicius, in Chapter 29, provides an insight into the process of industrialization in planned economies under the strong influence of the USSR. He argues that while there is extensive pollution and contamination, especially where Soviet military bases were located, the situation is not as bleak as is often suggested. Rinkevicius then goes on to focus on the results of an extensive survey of Lithuanian industry which aimed to identify the major stimuli and barriers to environmental management adoption. He attempts to address the key question of how painful it is for industry in the period of economic transition in Lithuania, asserting that the lack of information and economic interest among employees and the underdeveloped environmental legislation in the country are all key barriers to waste and pollution minimization in Lithuanian industry. Rinkevicius pinpoints the emergence of *perestroika* as a decisive catalyst in increasing the number of environmental activities in Lithuanian companies. He concludes that to achieve environmental policy changes environmental policy needs to be integrated and aligned with other government policies.

In Chapter 30, Vladimír Dodeš focuses on Czech industry and cleaner production projects. He asserts that Czech industry is facing the twin imperatives of the need to increase productivity and competitiveness and the need to improve environmental performance; however, he suggests that industry views these two imperatives as incompatible. Dodeš discusses the practical results of 24 cleaner production case studies drawn from a collaborative Czech–Norwegian project. The project identified substantial economic and environmental benefits and he argues that significant potential for economic and environmental improvements exists in all Czech industry. Dodeš considers the wider implications of the cleaner production projects for environmental management systems implementation in Czech industry. He concludes that enterprises can most effectively achieve the aims of their environmental policies by using cleaner production techniques and that environmental management systems are an important tool to make cleaner production implementation sustainable.

In the final chapter of this section, Gulbrand Wangen expands on Dodeš' conclusions by discussing the introduction of environmental management systems in combination with cleaner production. He draws on examples from Western Europe, Norway, and Central Europe, Hungary. Wangen argues that it is the current trends of internationalization of trade and the harmonization of standards which make environmental performance an increasingly important competitive factor. He describes the results of the Norwegian cleaner production approach which is set against a background of advance integrated pollution control. Wangen then discusses the development of quality and environmental management systems in a Hungarian company, pointing out that the company's management structure was bureaucratic with ill-defined responsibilities, both of which prevented the company from meeting the challenges it faced. He concludes that the enterprise management culture in CEE countries is still fairly authoritarian and as such new management approaches are not disseminated to the workers.

One of the reccurring themes in this section is the degree to which employees are involved in their company's environmental management. This theme did not feature as highly in the European industrial case studies discussed in the preceding section. The authors recognize the need for employee involvement in the achievement of performance improvements of an enterprise, but there is also agreement that fundamental and basic employee training is necessary to engage staff fully in the improvement process. The current lack of staff involvement, coupled with their lack of interest and poor communication between management and employees, all have serious implications for the successful implementation of environmental management systems and cleaner production projects. The management style in transition economies is typically characterized by the top-down rather than the bottom-up approach. In particular, the critical analysis of CEE countries shows a rigid, prescriptive and bureaucratic management style which disassociates the workforce and negates the possibility of employees contributing to environmental projects and therefore helping make them a success.

Another reccurring theme is the weakness of existing national environmental policy, legislation and regulators, which all conspire to give the wrong signals to industry. The section shows that political and social change is bringing about the need for enterprises to change. Management systems have a role to play in helping to structure the cultural change required in the business around formalized systems. The same systems will also help to make cleaner production projects sustainable. The theme of management culture and the lack of management capacity to address its businesses' environmental impacts are continued in the final section of this book, which focuses on the experiences of small and medium-sized enterprises implementing environmental management systems.

27
Environmental management systems for cleaner operations in South African mines

HAROLD NICHOLLS

INTRODUCTION

In South Africa there has been a concerted attempt by both the owners of mines and government authorities to reduce the impact of mining on the environment. Mine owners are required to produce an environmental management programme which will minimize the environmental impacts during all phases of their mining operations. This commitment is then enforced by law. In order to comply with this commitment it is necessary to ensure that the mining operations are monitored and audited regularly. This is achieved by putting in place an environmental management system.

Experience has shown that this is still not enough. Dedication by all the operational staff to achieve these commitments is needed. In other words, some form of environmental capacity building should be included so that operators take ownership of the commitments given by the mines' owners.

This chapter is a case study of the South African experience, showing how legally binding commitments, the implementation of environmental management systems, environmental capacity building and environmental awareness campaigns are all an integral part of an environmental management system for cleaner operations.

Environmental Management Systems and Cleaner Production, edited by R. Hillary.
© 1997 Anglo American Corporation of South Africa Ltd.

BINDING COMMITMENTS TO PROTECT THE ENVIRONMENT

Historical development of environmental management in South Africa

The Minerals Act 1991 sought to consolidate the various existing pieces of legislation on mine closure by requiring that a rehabilitation plan and programme be prepared for all mines and mineral prospecting sites.

Before 1991 numerous different laws and regulations were set in place which required the involvement of a large number of government departments. There were three main laws:

1. The regulations in terms of the Water Act 1956 addressed water pollution by setting standards with which effluents had to comply. In 1975 an arrangement known as the Fanie Botha Accord was concluded, in which the rights and obligations of mining companies in relation to water pollution-control measures after mine closure were clearly defined for mines that had closed before 1956.
2. The Mines and Work Act 1956 required that a rehabilitation plan and programme be submitted only in the case of opencast mines where more than certain tonnages of material were removed. The Inspector of Mines was required to call for rehabilitation planning in the case of other mines.
3. The Atmospheric Pollution Act 1965 required mines to adopt the best practical means for dust prevention. This legislation empowered the Minister to establish a Dust Control Levy Account, which resulted in many mines establishing trust funds in lieu of making contributions to the Levy Account. In time these trust funds developed into rehabilitation funds.

The Minerals Act 1991 was later amended and the rehabilitation plan and programme were replaced with an environmental management programme report (EMPR). Under the terms of the amended Act all mines and prospecting ventures are required to reapply for a mining authorization and an EMPR has to be prepared and submitted with the application to continue mining or prospecting.

The prime objectives of the EMPR are to develop and cost a mine closure plan and an environmental management programme. This mine closure plan has to satisfy the requirements of several government departments, including the Departments of Water Affairs and Forestry, Mineral and Energy Affairs, Health, Agriculture and Environment and Tourism. The EMPR is submitted to the Department of Mineral and Energy Affairs as lead agent, which distributes it to other departments. Once these departments approve the report it is up to the mining company to ensure that the commitment given therein is followed in all respects. If a mine satisfies the obligations set out in the EMPR, closure will become a formality and a closure certificate will be issued. The mining company's responsibility for the care of the environment will then have been completed and any future care will be the responsibility of the state. Since an EMPR is prepared many years before mine closure actually occurs, circumstances can change that will require amendments to the EMPR. Any amendments will again require the approval of all the government departments. Likewise, the costs of closure given in the EMPR will have to be re-evaluated at regular intervals and amended when necessary.

Environmental management programme report—document specifications

To make the structure of all EMPR consistent—and thus, easier for the authorities to evaluate—each mine is required to present its EMPR according to a format prescribed in the *Aide-mémoire*. This *Aide-mémoire*, which was developed by the Department of Mineral and Energy Affairs in collaboration with the mining industry, describes in detail what must be included in each section. The EMPR is structured as a logical process that a mining company follows to develop environmental management plans and programmes.

An EMPR is required for any mining operation, including prospecting and quarrying. The EMPR and *Aide-Mémoire* are dynamic documents that will require regular reviewing and updating as circumstances change.

Structure of EMPR for all mining operations

The EMPR is devised into six main parts, of which parts 1–4 provide background information and define the project:

- Part 1—Brief project description
- Part 2—Description of the pre-mining environment
- Part 3—Motivation for the proposed project
- Part 4—Detailed description of the proposed project
- Part 5—Environmental impact assessment
- Part 6—Environmental management programme

The environmental impact assessment (part 5) examines the impacts of the mining operation during the construction, operation, decommissioning, closure and post-closure phases. The objective is to highlight the magnitude of the impacts during each phase. For example, the impact could be slight, moderate or high and could have an effect on a local, regional, national or international level. Those impacts which are identified as being moderate or high are then ameliorated by the implementation of suitable management plans.

The environmental management commitment is given in part 6. Each management programme should provide a management goal and give details of the management objective, how it is to be implemented and who is responsible. Information on timing of the management programme and training requirements should also be provided. The methods described must adhere to the principle of best available technology not entailing excessive cost (BATNEEC). In addition, details of all monitoring programmes must be included so that all concerned can be certain that the measures in place to protect the environment are adequate and effective. The environmental management programme is of great importance since it must provide sufficient information to satisfy the authorities that the impacts identified as significant are being properly ameliorated.

Examples of a number of management programmes are given later.

Part 6 must also include the cost estimates associated with the management programme. The cash flow that is required throughout the life of the mine to manage the environment and to ensure satisfactory mine closure is illustrated in Table 27.1.

The cash flow forms the basis for an environmental fund that is set up to ensure that sufficient financial resources are available for mine closure. Currently, the law dictates that equal annual payments are required up to decommissioning. That no flexibility is allowed in these contributions, thereby eliminating the possibility for innovative financing, is possibly a shortcoming of the law. Contributions to the fund are not taxable, so their amount must be approved by the Receiver of Revenue. An example of such a fund is given in Table 27.1. The cost of closure should be reassessed regularly to allow for changing circumstances and increasing costs. The interest earned by the fund can, however, be used to offset the effects of inflation and hence reduce the need for frequent updating of the estimated cost of closure.

Any amendments that are made to the EMPR from the time when it is first approved to the time of the final closure of the mine are included under a separate part of the EMPR.

Table 27.1 Example of contributions and expenditure in an environmental trust fund (10 000 000 Rand)

Year	Expenditure	Contributions	Income	Fund value
1994		1.15	0.00	1.150
1995		1.15	0.115	2.415
1996		1.15	0.242	3.807
1997		1.15	0.381	5.337
1998		1.15	0.534	7.021
1999	0.5	1.15	0.652	8.323
2000	2.9	1.15	0.542	7.115
2001	0.1	1.15	0.702	8.867
2002	2.8	1.15	0.607	7.823
2203	4.7	1.15	0.312	4.486
2204	4.5	1.15	0.009	1.244
2005	1.1	1.15	0.014	1.309
2006	1.0	0.90	0.031	1.240
2007	0.9		0.034	0.374
2008	0.3		0.007	0.081

Simplified EMPR for small-scale mining and prospecting operations

The EMPR as described above is a comprehensive, in-depth document. As it is unreasonable to expect a small mining operation to produce such an in-depth report, the Department of Mineral and Energy Affairs has produced an abridged *Aide-Mémoire* in the form of a questionnaire. The operators of small mines will have to complete this questionnaire and, by so doing, give an undertaking on how they will close their operations.

SUPPLEMENTARY DOCUMENTS TO ASSIST IN PREPARATION AND IMPLEMENTATION OF AN EMPR

To assist mines further in the preparation and implementation of an EMPR, a series of policies and guidelines are currently being prepared under the headings *Minerals Act 1991: Policy on financial provision* and *Policy on mine closure*. The objectives of each of these are summarized below.

Minerals Act 1991: policy on financial provision

The main objectives of the financial policy are to assist the authorities in confirming that mines can make the necessary provision for the execution of an approved EMPR; to identify ways in which such provision may be made; to identify the circumstances in which the state will assume responsibility for the management of land where there is no legally responsible person; and to identify the circumstances in which the state will, in other cases, assume responsibility for obligations relating to the environmental management of land disturbed by mining operations.

Policy on mine closure

The policy on mine closure is still being determined through negotiation with the Chamber of Mines, which represents the mining companies and the various government departments. The principles on which this policy is being formulated are: on fulfilment of the commitments given in the EMPR a mining company shall be entitled to receive a closure certificate; prior to the issuing of a closure certificate a monitoring period shall elapse from the date of cessation of the mining operations; if the post-closure maintenance risk is low, an unconditional closure certificate will be issued and the state will assume liability; if the post-closure maintenance risk is high, the mining company will need to make provision for post-closure maintenance costs; and the closure requirements shall be based on BATNEEC.

THE IMPLEMENTATION OF AN ENVIRONMENTAL MANAGEMENT SYSTEM

In the earlier chapters of an EMPR an environmental impact assessment is undertaken on the mining operations. In a later chapter the legal aspect is considered. Hence the EMPR also contains some of the essential components of an EMS. There are, however, some items in addition to the EMPR that are still required to complete the EMS. These are short-term management plans and monitoring and auditing requirements. The documents which are produced within an EMS are given below:

- Environmental policy
- Impact register
- Legal register
- Interested and affected party register
- Long-term management plan
- Short-term management plan
- Monitoring register
- Internal audit protocol

Figure 27.1 depicts how the EMPR fits into the ISO 14001 environmental management system.

Any EMS can be divided into three main categories:

1. *Planning:* the objectives and targets which are set in place up front to drive the management system.
2. *Implementation:* the operating management plans which ensure that the various activities involved in mining take cognizance of the objectives and targets set in place.
3. *Measurement and evaluation:* the monitoring, auditing and reviewing functions which ensure that the objective and targets are met or complied with.

Each of the above is depicted in Figure 27.2.

Figure 27.2 clearly sets out three different functions of the EMS. These should be staffed separately. (In the case of small operations this may not be possible.)

- **Planning aspects:** The person in charge of this function is responsible for setting up the EMPR. The EMPR addresses the legal requirements and includes an impact assessment from which management goals and objectives are set. There are, however, two additional aspects that must also be considered. First, the company's environmental policy should give direction to the environmental management that has been set in place. Second the concerns and perceptions of interested and affected parties must also be managed. These concerns and perceptions are evaluated and where applicable add to the commitments given in the EMPR.
- **Implementation aspects:** The miners and operators are responsible for managing their activities within the constraints of the commitments identified above. For

CLEANER OPERATIONS IN SOUTH AFRICAN MINES

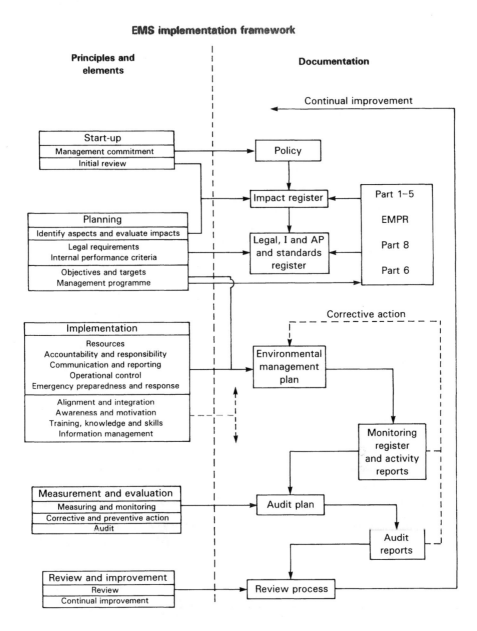

Figure 27.1 ISO 14001 environmental management system including an EMPR

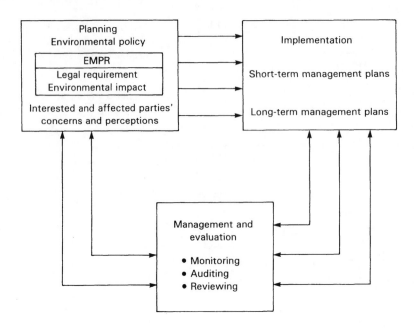

Figure 27.2 Three different functions of an EMS

this management to be effective they must have a clear understanding of what the specific management needs are.
- **Measurement and evaluation aspects**: This duty is separated from the operational duties in order objectively to monitor and evaluate the operation's programme. It also reviews the various documentation compiled under all three sections to ensure that these are kept up to date and that there is some continuity.

With the management system in place and binding commitments given in the EMPR to minimize environmental damage, the only outstanding element of the EMS is to ensure that the operators clearly understand their obligations.

CAPACITY BUILDING AND ENVIRONMENTAL AWARENESS

In an attempt to gain the support of all personnel, interactive workshops are being held on some mines. These workshops are attended by middle and senior managers who are involved in the implementation of the environmental management system. The aim and objectives of each workshop are briefly discussed below.

Workshop 1

In this workshop the mining operation is divided into its main components. For example:

- mining
- ore processing
- engineering services

The main activities within each component of the operation are then identified. For example:

- Mining:
  - removal of overburden
  - blasting
  - removal of ore
  - stockpiling ore

Once all the activities have been identified the various linkages between each activity and the environments are identified. These links are the environmental aspects. For example:

- solid waste production
- gaseous emissions
- dust emission
- liquid waste
- vibrations
- disturbance to land

For ease of understanding each activity and associated environmental aspects are incorporated into a matrix, as shown in Table 27.2. The preparation of this matrix will encourage the operator to think about how each of his other activities is linked to the environment. For many operators this linkage has never been a consideration.

With a clear understanding of the operations the next step is to list all the environments and to ask the operators to assess how each environment will be affected by each activity. This is achieved by extending the above matrix to include environments. An example of an assessed matrix is given in Table 27.2.

A great deal of capacity building is achieved in this workshop. Experience has shown that this is the first time many operators actually identify what damage their activities are afflicting on the environment. The cause of this damage also becomes identifiable.

Workshop 2

In this workshop management plans are developed for those activities which have significant impacts on the environments. A code of practice must be developed for

Table 27.2 A matrix relating activities and aspects to impacts on the environment

Task	Activity	Aspects				Environments										
		Air	Solid waste	Liquid waste	Vibration	Geology	Topography	Soils	Land capability	Land use	Vegetation	Animal life	Surface water	Groundwater	Noise	Air quality
Mining	Removal and transport overburden	*	*					•		•	•	•			•	•
	Blasting removal of ore	*	*		*										•	•
	Transport of ore	*													•	•
	Stockpiling of ore	*				•	•							•		•

each activity that has been shown to have had a significant impact. The main advantage of the matrix is that the operator focuses on only those environments that are affected. An example of management plan identification is given in Table 27.3.

The framework for a typical activity code of practice is given below:

- identify responsible person;
- describe activity briefly;
- list management objectives (obtained from impact matrices);
- list environmental concerns (also obtained from impact matrices);
- describe management actions;
- describe reporting;
- identify training needs.

These codes of practice are developed by the operators and are then reviewed by an environmentalist. The end product is the short-term management plans.

The long-term management plan will have been developed in the EMPR and discussed earlier.

Since the operators have developed their own management codes of practice with the full knowledge of why these are necessary, the management codes are more likely to be taken seriously and implemented correctly.

Workshop 3

Personnel involved in policing the effectiveness of the activity management plans will attend this workshop. It examines monitoring plans using the matrix developed in the first workshop.

The monitoring aspect associated with the matrix is given in Table 4.

The monitoring plans for a specific environment will monitor the effects of the various activities on that environment. A typical contents of a monitoring programme is given below:

- responsible person
- objective
- relevant activities
- monitoring method
- communicating results
- annual reporting and review
- supporting information

Codes of practice are more likely to be taken seriously and implemented correctly.

Workshop 4

The aim of this workshop is to pull together all that has been learned in the previous three workshops with other items found in the EMS such as the legal procedures, interested and affected parties register, the environmental policy and the EMPR.

Table 27.3 Impact matrix relating activities to impacts

Task	Activity	Aspects				Environments										
		Air	Solid waste	Liquid waste	Vibration	Geology	Topography	Soils	Land capability	Land use	Vegetation	Animal life	Surface water	Groundwater	Noise	Air quality
Mining	Removal and transport overburden	*						○		○	○	○			○	○
	Blasting removal of ore	*	*		*										●	●
	Transport of ore	*													●	●
	Stockpiling of ore	*				●	●							●		●

○ Issues to be addressed in activity management plans for removal and transport overburden

Table 27.4 Matrix depicting monitoring requirements

Task	Activity	Aspects				Environments									
		Air	Solid waste	Liquid waste	Vibration	Geology	Topography	Soils	Land capability	Land use	Vegetation	Animal life	Surface water	Groundwater	Noise	Air quality
Mining	Removal and transport overburden	*	*					●		●	●	●			○	●
	Blasting removal of ore	*	*		*										○	●
	Transport of ore	*													○	●
	Stockpiling of ore	*				●	●							●		●

○ Monitoring requirement for noise

Legal procedures

A computerized legal procedure has been developed which relates all the activities identified in the first workshop to the laws and regulations of the country. The legal procedures enable the operator to identify at a glance what are the legal responsibilities and permit requirements. The actual permit can be scanned into this legal register. Furthermore, where necessary a commentary can be added to clarify exactly what is required by the law.

An example of this legal matrix is given in Table 27.5.

Interested and affected parties register

The interested and affected parties register is a computerized database in which all communications with interested and affected parties are logged. A program such as *PIMS* (Public Involvement Management System), developed at the University of Cape Town, can be used for this purpose.

Environmental policy

The environmental policy developed by mine management will have taken cognizance of the activities identified in the first workshop. It will clearly set out what is expected of mine personnel and how the mine intends to manage the environment effectively.

Auditing and review of the EMS and the importance of a correctly structured audit protocol are also discussed at this last workshop.

At the end of the four workshops senior and middle management should have a clearer understanding of what drives environmental management, what impacts the various activities have on the environment, what the legal obligations are and finally an appreciation of the importance of the short- and long-term management and monitoring plans.

CONCLUSION

In order to achieve a cleaner operation the following considerations have equal importance:

- binding commitments;
- an environmental management system;
- capacity building in the workplace so that operators are committed to the success of the EMS.

With all the above in place cleaner production will be achieved.

NOTE

It is acknowledged that the basis of this chapter has already been used in material published by the Anglo American Corporation of South Africa Limited ©, and is reproduced with permission.

Table 27.5 An example of a section of the legal register

Task	Activity	Minerals Act			Water Act		
		Construction	Operation	Closure	Construction	Operation	Closure
Mining	Removal and transport of overburden	●					
	Blasting and ore removal		●				
	Transport or ore		●		●		
	Stockpiling of ore		●				

[Pop out window with commentary on this section of the act.]

28
Environmental management initiatives in the Brazilian oil industry

SERGIO PINTO AMARAL

INTRODUCTION

Since 1953, in Brazil all activities relating to oil exploration, drilling, production, transportation and refining have been the monopoly of the federal government. The company responsible for acting in the government's name to conduct these activities is Petróleo Brasileiro SA, Petrobras. The marketing of oil products is not part of this monopoly, being left to national and international corporations. The national production of oil is around 850 000 barrels per day, 70% of which comes from off-shore production fields located approximately 80 km off the coast in a region called the Campos Basin in the State of Rio de Janeiro.

At present the company's refining capacity is 1 572 000 barrels of oil per day and the national daily consumption presently stands at around 1 350 000 barrels. Petrobras has two subsidiaries, one for distributing its products and the other dedicated to its international activities.

The company is active in most Brazilian states and in several countries abroad in the different areas of the oil industry, notably in exploitation and marketing of oil products. It is recognized internationally for its technology in off-shore oil exploration in depths of water over 1000 metres. There are about 42 000 employees and over 40 operational units spread throughout Brazil, such as districts for exploration, drilling and production, oil terminals and refineries.

Environmental Management Systems and Cleaner Production, edited by R. Hillary.
© 1997 John Wiley & Sons Ltd.

The company has a written environmental policy and the environment, quality and industrial safety functions are managed jointly. However, the responsibility for the operationalization of these functions belongs to the different sectors and line activities, such as project, operation and maintenance.

Furthermore, since mid-1991 the company has been implementing a total quality management (TQM) programme with the objective of continuously improving its activities. At present, Petrobras is assessing its performance based on the criteria of excellence for the National Quality Award (Fundação para o Prêmio Nacional da Qualidade 1995), which emulates the US Malcolm Baldrige National Quality Award.

This chapter, which contains some ideas and information featured in the Master's thesis "Environmental auditing in the oil industry" developed by the author (Amaral 1992), presents the link in Brazil's oil industry between the application of environmental management systems and the promotion of cleaner products and processes. Discussion will deal with Brazilian and international laws, regulations and standards with regard to environmental management and auditing systems; the environmental management system in place in Petrobras; the initiatives taken by the company towards delivering cleaner products and processes; and the programmes introduced over the last few years in the area of environmental promotion, training and raising of awareness.

BRAZILIAN AND INTERNATIONAL LAWS, REGULATIONS AND STANDARDS ON ENVIRONMENTAL MANAGEMENT AND AUDITING

In March 1993, the International Organization for Standardization (ISO) set up the Technical Committee TC 207, Environmental Management Systems. The purpose of this committee is to draw up a set of norms on environmental management that is named the ISO 14000 series of standards, as can be seen in Figure 28.1. It is expected to oversee the publication of around 18 standards in various areas of environmental management.

With respect to Europe, in January 1994 the British Standards Institution (BSI) published the revised version of *BS 7750—Specification for Environmental Management Systems* (BSI 1994). This standard is being used as a basis by the ISO's TC 207 SCI in matters concerning environmental management systems.

Environmental Management

ISO 14000

| SC-1 Environmental Management Systems | SC-2 Environmental Auditing | SC-3 Environmental Labelling | SC-4 Environmental Performance Evaluation | SC-5 Life-Cycle Assessment | SC-6 Terms and Definitions | WG-1 Environmental Aspects in Product Standards |

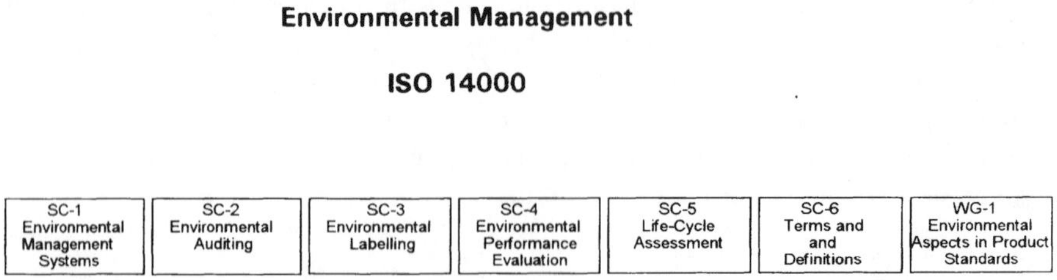

Figure 28.1 ISO 14000 series of standards (Source: adapted from Harmon 1994)

Also with regard to environmental auditing, the European Union (EU) published the *Community Eco-management and Audit Scheme—EMAS* (Commission of the European Communities 1993) for companies to undertake voluntary environmental audits. This scheme came into force in Spring 1995. A number of European enterprises are qualifying for certification through BS 7750 and EMAS.

To date, in Brazil, no standard or regulation on environmental management exists. The tendency is for companies voluntarily to use the international norms drawn up by the ISO. Accordingly, a number of Brazilian companies that operate overseas, including Petrobras, set up a working group together with the Brazilian Association of Technical Norms (ABNT) in order to comment on the deliberations of TC 207.

The purpose of this group, called GANA (Supporting Group on Environmental Standardization) is to assist the ABNT in analysing all the draft standards on environmental management being drawn up by ISO so that the ABNT, which represents Brazil before the ISO, can announce the country's official position to that organization.

As for environmental audits, the Chamber of Representatives of three Brazilian states, namely Rio de Janeiro (Law No. 1898 of 26/11/91), Minas Gerais (Law No. 10627 of 16/01/92) and Espírito Santo (Law No. 4802 of 02/08/93), passed laws making periodical environmental audits obligatory for certain industrial activities, such as the production and processing of oil, chemicals and steel. There also exists a bill at the federal level (No. 3160 of 26/08/92), now under discussion in the Brazilian Congress, proposing to make it obligatory for certain business activities to undertake environmental audits. This proposal is more wide ranging than the state laws and is based on the European Union's EMAS.

What can be gathered from the emerging Brazilian legislation on the obligatory nature of environmental audits is that the legislators, concerned about the weakening of the official agencies for environmental control, are trying to transfer responsibility for control to the production activities.

THE ENVIRONMENTAL MANAGEMENT SYSTEM IN PLACE IN PETROBRAS

Petrobras has been concerned with environmental issues since the beginning of its operations. In 1974—only two years after the United Nations Conference on Human Environment held in Stockholm, Sweden—the company published its first "Industrial Protection Policy", containing the ideas and knowledge of that time.

In 1985 the "General Principles of Protection" were defined. The proposed guidelines anticipated the occurrence of environmental problems and lent the same priority to environmental issues as to production, and stressed that environmental responsibilities should be a management concern.

In 1989, Petrobras drew up an environmental policy composed of general guidelines for environmental management. Priority has been given to preventive strategies; suitability of technological and material resources; training and commitment of the human resources involved; concern for economic and social development without jeopardizing the society's quality of life; intensified use of technologies of low-polluting impact; and creation/implementation of local and regional contingency plans.

This policy was revised in January 1996, in order to include some novel issues which are emerging in this decade, such as respect for the principles of sustainable development and the implementation of environmental management systems and auditing.

The company's environmental structure is composed of a series of sectors where about 200 professionals from different backgrounds (engineers of various specializations, geologists, biologists, chemists, lawyers, technicians and specialized operators) work in departments and activities related to environmental matters. These professionals work in the various states of Brazil where the company is active.

The Superintendency for the Environment, Quality and Industrial Safety (SUSEMA) is the central body that formulates the environmental policy, guidelines and standards for the whole company, as well as articulating and controlling the different environmental programmes being developed by Petrobras. Each year it invests about 7–10% of the company's overall budget in programmes and projects related to the environment. At present, this amounts to something in the order of US$150–200 million per year. Some examples of these environmental projects and programmes are waste water treatment works; programmes to improve fuel quality; systems for treatment and disposal of solid waste; environmental monitoring programmes; and purchase of equipment for fighting oil pollution at sea.

In addition to this, the company's staff is implementing an "environmental and industrial safety corporate auditing programme". The aim of this programme, to which the top management of the company is fully committed, is to ascertain whether the pertinent functions are being properly managed and whether the company's policy and the legislation with respect to such matters are being complied with.

In principle the programme is to be internal with two levels of auditing, one corporate and the other departmental. In August 1994, the company concluded the training of 100 professionals within the company (50 in the area of environment and 50 in industrial safety) in auditing techniques and skills.

It is also preparing to meet the forthcoming Brazilian legislation on environmental auditing and, in the future, the company will fit its environmental management system to international models, in accordance with the future series of environmental standards ISO 14000.

PETROBRAS'S ENVIRONMENTAL INITIATIVES RELATING TO CLEANER PRODUCTS AND PROCESSES

In the last few years, the increase in society's environmental awareness, the action of governmental and non-governmental organizations engaged in environmental control, and the sense of environmental responsibility on the part of company managers and professionals have enjoyed remarkable growth. Consumption of environmentally friendly products, recycling initiatives and the growing demands of consumers who seek information on how products are made are creating a "green market" that modern, forward-looking companies have to take into consideration.

In view of this, in the last few years Petrobras has been searching for environmentally sound technologies and products. A description follows of the main initiatives within the company's environmental management system with a view to delivering cleaner products and processes.

Addition of ethyl-alcohol to gasoline

The addition of oxygenated compounds to gasoline is becoming an alternative to the use of lead compounds as anti-knock compounds for gasoline. In order to eliminate the pollution caused by lead from Brazil's large urban centres, and because the country boasts an ethyl-alcohol production programme based on sugarcane (Pro-álcool), Petrobras and the Brazilian government decided in 1989 to substitute the addition of tetra-ethyl lead to Brazilian gasoline with ethy-alcohol. At present all gasoline in the country must contain 22% ethyl-alcohol.

Use of low-sulphur diesel in large urban centres

There has been a substantial increase in the use of diesel vehicles in Brazil's urban centres. In 1992, in order to improve the quality of air in these cities, Petrobras signed a letter of intent with the Brazilian Institute for the Environment and Renewable Natural Resources (IBAMA) to supply diesel with 0.5% sulphur to nine Brazilian cities. Within this programme to reduce the sulphur content in diesel from 0.7% to 0.5%, the company will install five hydrodesulphurization (HDT) units at its refineries by the year 2000, at a cost of approximately US$1.25 billion.

Vehicle-emission control laboratory

In order to obtain a better identification of the polluting emissions of national and foreign vehicles that use the fuels produced by Petrobras, the company has invested US$5.5 million in building a Vehicle-Emission Control Laboratory at its Research and Development Centre in Rio de Janeiro. Inaugurated in 1992, this laboratory performs different tests to measure CO, SO_2 and other pollutants in diesel and gasoline vehicles and is used for better product development.

The use of new chartered-in ships for oil transportation

In the last three years, the company has been improving its ship-freighting procedures in Brazil and abroad. One of the new measures is the reduction of the average age of chartered-in ships, down from 20 to 12 years. The main objective of such procedures is to operate with greater reliability and reduce the likelihood of an accident, thereby avoiding an oil spill at sea.

The use of natural gas as alternative fuel in homes and industries

In order to provide cleaner fuel for homes and industries, the company has been building gas pipelines and branches for many users in an attempt to increase the supply of natural gas to homes and industries as a substitute for traditionally supplied naphtha and fuel oil. Petrobras has been using this procedure mainly in the states of Rio de Janeiro and São Paulo, utilizing the gas from the Campos Basin, since these states are the most densely populated and most critical in terms of air pollution. In the near future, the company plans further to increase the supply to the states in the centre-west region and other states in the south-east and south with natural gas from Bolivia. This will be made possible by the recently signed agreement between the governments of Brazil and Bolivia for the construction of a gas pipeline joining the two countries.

ENVIRONMENTAL PROMOTION, AWARENESS RAISING AND TRAINING

Recently Petrobras has sponsored and supported a series of environmental conservation projects under external initiatives. It is company policy to have its image associated with environment-linked initiatives. Table 28.1 presents the 10 leading environmental preservation and conservation projects supported by the company in several states throughout the country.

As regards education and raising awareness on environmental issues, there has been an increase in practical initiatives in the company, such as recycling and reuse programmes. For instance, the different offices in Petrobras have litter-recycling schemes. In the refineries there are ever more selective collection programmes and recycling of materials such as paper, glass, plastic and metal. The money collected from the sale of recycled material is normally allocated to educational programmes supported by the company. Petrobras has also been experimenting with the use of oily sludges in the making of ceramic materials—a good example of how to use a residue of the oil industry in other productive sectors (Amaral 1991).

In the last few years Petrobras has held a series of courses and seminars for its professionals in the following areas: environmental management and auditing, environmental law, environmental impact assessment, waste water treatment and solid waste disposal, air pollution control, effects of pollutants on eco-systems, and oil-spill fighting at sea.

CONCLUSIONS

Brazil's oil industry has made great efforts in the last 20 years in the areas of environmental preservation, prevention and control. Given the high degree of

Table 28.1 Environment-linked external projects supported and sponsored by Petrobras in Brazilian states

- Preservation of sea turtles (Espírito Santo, Bahia, Sergipe)
- Preservation of cetaceans (Rio de Janeiro)
- Adoption of the biological reserve in the Arvoredo archipelago (Santa Catarina)
- Adoption of the archaeological site in Lajedo de Soledade (Rio Grande do Norte)
- Recovery and preservation of the world's largest cashew plantation (Rio Grande do Norte)
- Agreement between Petrobras and the University of the Amazon for the study of the tropical forest, where the company operates (Amazonas)
- Adoption of the environmental protection area in Monte Pascoal (Bahia)
- Young ecologists' brigade on Ilha Grande (Rio de Janeiro)
- Monitoring of the lakes in the vicinity of the Campos Basin (Rio de Janeiro)
- Eco-development Institute in the Ilha Grande Bay (Rio de Janeiro)

Source: Petrobras

awareness on the part of the company's managers and employees of matters concerning the environment, these activities are expected to continue as priority issues within Petrobras's philosophy.

Furthermore, environmental demands from society, consumers, governmental and non-governmental organizations concerned with environmental preservation and control will continue to grow, and this will lead the company to remain constantly on the alert regarding environmental matters. On the other hand, the future series of international standards for environmental management (ISO 14000) will also demand that modern corporations follow towards cleaner products and processes and environmentally more sound technologies.

REFERENCES

Amaral, S.P. (1991) "Oily wastes application in ceramic materials manufacturing", *Water Science and Technology*, Vol. 24, No. 12, pp. 165–176.

Amaral, S.P. (1992) "Environmental auditing in the oil industry". A report submitted in partial fulfilment of the requirements for the MSc degree and the DIC, Centre for Environmental Technology, Imperial College of Science, Technology and Medicine, London, UK.

Bill No. 3160 of 26/08/92, concerning environmental audits in institutions whose activities have an impact on the environment, *Câmara Federal dos Deputados*, Brazil.

British Standards Institution (1994) *BS 7750: Specification for Environmental Management Systems*, London, British Standards Institution.

Commission of the European Communities (1993) "Council Regulation (EEC) No. 1836 93 of June 1993 allowing participation by companies in the industrial sector in a Community eco-management and audit scheme", *Official Journal of the European Communities*, 10 July.

Fundação pare o Prêmio Nacional da Qualidade (1995) *Critérios de Excelência—O Estado da Arte da Gestão pela Qualidade Total*, São Paulo, Brazil, Fundação pare o Premio Nacional da Qualidade.

Harmon, M. (1994) "First there was ISO 9000, now there's ISO 14000", *Quality Digest*, July, pp. 25–31.

Law No. 1898 of 26/11/91, concerning environmental audits, *Diário Oficial do Estado do Rio de Janeiro*, Brazil, Wednesday 27 November 1991, Year XVII, No. 228, Part 1.

Law No. 10627 of 16/01 /92, concerning environmental audits and other measures, *Diário Oficial do Estado de Minas Gerais*, Brazil.

Law No. 4802 of 02/08/93, concerning environmental audits and other measures, *Assembléia Legislativa do Estado do Espírito Santo*, Brazil.

29
The greening of Lithuanian industry: past and present

LEONARDAS RINKEVICIUS

INTRODUCTION

The concepts of environmental management, eco-auditing, pollution prevention and cleaner production are gradually becoming key words (sometimes buzzwords) for industrial decision makers and business strategists. The environmental dimension was not at the top of the list of concerns guiding industrial decision making in the 1960s and 1970s, but since the mid-1980s it has acquired much greater importance (Fisher and Schot 1993). Pollution prevention and waste minimization are increasingly recognized as pathways leading not only to environmental improvements, but also to economic benefits—e.g. production efficiency, good payback of preventive measures, "green" image for the companies in society, reduced waste disposal and treatment costs, better preparedness for future laws and regulations which are likely to become more and more stringent.

These changes and challenges are characteristic of leading industrial countries which have advanced innovation systems and maintain a sufficient technological level to allow them to sustain a competitive edge. What about countries which are behind the leaders but striving to catch up? In particular, what about Eastern European industry?

This chapter briefly sketches out environmental management in Lithuanian industry. It might be argued that Lithuania is somewhere in the upper middle level as compared with its other Eastern European neighbours in terms of general trends of

---

Environmental Management Systems and Cleaner Production, edited by R. Hillary.
© 1997 John Wiley & Sons Ltd.

socio-economic transition and particular stories of success and failure of industrial restructuring and incorporating environmental criteria in this process.

PROFILE OF LITHUANIAN INDUSTRY AND ENVIRONMENT: PAST AND PRESENT

Black smoke coming from chimneys, heavily polluted rivers and lakes, contaminated soil, unskilled and short-termist management are the typical attributes of the environmental landscape of Eastern Europe drawn by various (often Western) observers. Does this picture correspond to reality? The answer is both yes and no.

Yes, because there are a number of places where there is heavy pollution and a neglectful attitude to the environment. On the territory of perhaps all Central and Eastern European (CEE) countries there still remain contaminated areas where Soviet military bases have been located. Due to the subsidized type of planned economy the prices for energy, water and raw materials did not change for a long time, whereas the Western world was undergoing significant techno-economic changes often referred to as a decline of the Fourth Kondratiev cycle. The latter changes, induced to a large extent by two energy crises in 1973 and 1979, encouraged Organization for Economic Co-operation and Development (OECD) countries to undertake serious measures in technological restructuring aimed at significant reduction of energy consumption. Western industry was "rediscovering" industrial efficiency (Jamison et al. 1994), and gradually going beyond it.

Nevertheless, the world energy crises and other changes did not affect the internal market of the Soviet bloc and did not stimulate industries to seek significant technological and management renovation. The prevailing techno-economic paradigm institutionalized in the 1950s and 1960s continued. The technological trajectories and regimes (Nelson and Winter 1977), as well as management structure and style, were based on industrial approaches that were characteristic of the Third and upswing of the Fourth Kondratiev cycle, due to the material and energy intensity of production (Rinkevicius 1993; Brundenius 1995). Industrial practice did not internalize the growing awareness that there is "no free lunch", and that the costs and implications of waste and pollution have to be borne sooner or later.

It would be far from the truth to maintain that neglectful attitudes towards the natural environment were prevailing in the set of values and norms common to the "Soviet people". While the Soviet system often imposed dramatic limitations on what people could do as citizens, as consumers or as employees, industrial and environmental innovation did take place. It comprised several types of important developments.

The first type of innovation, which originated in the USSR (since about the mid-1970s) and was well known in Lithuania, is usually referred to as "low- and non-waste technology". It attracted a number of academics but barely reached the stage of implementation in industry. It was aimed to design industrial systems as a chain of production inputs and outputs which would close the loop, similarly the natural ecosystems, and thus any waste or pollution would be prevented from entering the natural environment. Such an approach has recently been "reinvented" in the Western world under the label "industrial ecology", and even has practical examples such as an industrial and municipal network (symbiosis) in the town of Kalundborg in Denmark and attempts to initiate so-called industrial ecology parks in New Jersey, USA or Burnside, Canada (Wallner and Fresner 1995).

Probably the main difference between non-waste technology and industrial ecology is its economic dimension. The former concept, embedded in the system of a centrally planned economy, primarily focused on the technical and ecological sphere of change. By contrast, the concept of industrial ecology (similarly to the concept of pollution prevention/cleaner production which can be viewed as its predecessor on a single-company basis) emphasizes that closing the loop should be economically viable and cost-effective. Moreover, the concept of industrial ecology implies a change which paves its way through a broad network on a regional or municipal level involving a variety of actors and stakeholders. In this regard, industrial ecology can be viewed as part of the broader concept and theory of "ecological modernization". In any case, the heritage of the "Soviet school of non-waste technology" provides fertile ground for further development of more advanced concepts, theories and practical approaches to industry and the environment.

Another kind of environmentally friendly innovation comes from the activities of so-called rationalizers, those people (usually engineers on the shopfloor) who, through their enthusiasm and creativity, contributed a vast number of incremental innovations in industry (for definition and description of incremental innovations see, for example, Freeman and Perez 1988, pp. 45–6). These innovations often led to improved production efficiency and resource/energy saving. Those people were often rewarded financially according to centrally set instructions and rates issued by the State Committee on Inventions and Innovation in Moscow, and usually their contributions were publicized in plant newspapers and awards were given in terms of diplomas, pictures on the bulletin board of honourable workers and other similar ways. The number of proposed and implemented rationalization options was one of the indicators according to which companies' achievements in accomplishing five-year plans were measured. In every former Soviet republic, including Lithuania, there existed a "society (association) of rationalizers" with local organizations in all the bigger towns. Unfortunately, the rationalization activity was poorly documented and in-depth research into this realm of incremental innovation is needed in order to estimate how significant the achievements were in terms of reduced waste and conservation of resources.

The third area of environmental awareness and innovation was nature conservation (and historical/cultural heritage preservation) clubs and societies. In Lithuania, these were often institutionalized within tourist clubs and folk music groups (usually at universities and research institutes). In 1987 and especially 1988, these social settings were dramatically transformed and made important contributions to forming a "green" movement. This movement was like a large monitor which was screening and exhibiting environmental mismanagement in the former Soviet economy. At the same time, this type of critique, and actors who articulated it, played an extremely important role as an awakening and mobilizing agency in the movement that opened up the ways for reestablishing national independence in particular CEE countries, e.g. Lithuania (Rinkevicius 1994).

The environmental awareness of people in the former Soviet campus is well reflected in environmental laws and regulations set at that time. Many CEE countries had enacted (and often still have) environmental standards and maximum allowed emission concentrations that are much more stringent than those in the West. But an important question is how the compliance with those strict norms and standards was managed, or rather mismanaged.

Generally, it might be argued that the Eastern European environmental situation is not as bleak as is often suggested or imagined. On the contrary, the fact is often overlooked that particular CEE countries, including Lithuania, are not among the leading "polluters" if taken in a cross-country comparison. For instance, the emissions per capita of SO_2 and NO_X in many European countries are much higher than in Lithuania. Statistics of the period of changes in CEE countries bear witness to the fact that emissions of NO_X per capita in Denmark, Finland or Germany (50–60 kg per capita in 1989) were twice as high as for Lithuania (e.g. 25 kg per capita). Comparing the per capita emissions of SO_2, the figure for Lithuania (60 kg per capita) is at least two or three times lower than those for Estonia, Poland or former Czechoslovakia (Halsnaes and Morthorst 1990). However, the ratio of waste and emissions per unit of produced good in many Lithuanian companies is far above the respective figures of Western industries. Moreover, the load of pollution in certain geographic areas is heavy enough seriously to harm people and the environment.

THE RESEARCH APPROACH AND SOME FINDINGS

The research project aimed to reveal major stimuli and barriers for environmental management in Lithuania.

In total, 620 companies were surveyed, representing all major sectors of Lithuanian industry: machine tool building, chemical, electrical and electronics, textiles, tannery, wood processing and furniture manufacturing, food processing, construction, transportation, energy and water supply. The main respondents were deputy directors and environmental managers or engineers of the companies, while the whole list of those who answered the questionnaire comprises 97 different posts and corporate functions. The response rate to the survey is relatively high—approximately 80%.

A total of 319 companies answered the question, "In which year did the company begin activities in the field of waste and pollution reduction?". There is no reason to believe that the rest of the sample which did not respond to this question have anything in common and that there is any pattern of behaviour within this group of companies. Thus it can be maintained that the answers of 319 companies are statis-

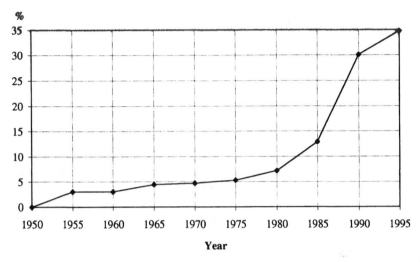

Figure 29.1 Percentage of companies which started environmental activities in different years

Table 29.1 Percentage of companies which started environmental activities in different years

Years	1950–55	1956–60	1961–65	1966–70	1971–75	1976–80	1981–85	1986–90	1990–95
% of companies	0.3	0.3	4.4	4.7	5.3	7.2	12.9	30.1	34.8

tically significant to provide a view on the time dispersion of the initiation of environmental activities in Lithuanian industry (see Figure 29.1 and Table 29.1).

Not surprisingly, there are very few companies that report the beginning of their environmental activities as taking place in the 1950s. In the 1960s, there is a tendency for a minor increase in the number of companies involved in waste and pollution reduction. The same is characteristic of the 1970s. It might be expected that, after the first Earth Day in 1970 and the UN conference on the Human Environment held in Stockholm in 1972, there should have been a more significant response both in terms of establishing and strengthening environmental authorities and simultaneously initiating environmental activities in particular industrial enterprises. Nevertheless, the survey results show that in Lithuania this was not the case, and a pattern of slow and steady development characteristic of the 1950s and 1960s continued till the mid-1980s.

There are several possible explanations. First, a greater industrial environmental response would have required an order from the government or even the Communist party, which was a typical pattern in the former USSR. Since there were no significant changes or signals "at the top", the majority of companies kept on doing business as usual.

On the other hand, it is unlikely that many people who filled out the survey questionnaire worked in the same company or in a similar position in the early 1970s. Therefore, it is not surprising that respondents have identified those years during which they themselves played a role, or at least have better knowledge of that period (e.g. the late 1980s). However, there are still a few people who participated in launching environmental projects and programmes at companies in the 1970s and early 1980s, or have memories of such events. Undoubtedly, this area is underresearched, and it needs further investigation by historians, sociologists and engineers.

The period from the second half of the 1980s until the present can be characterized as the time of the most notable shift towards waste and pollution reduction in Lithuanian industry. It is important to draw a parallel between this shift and the shift in corporate responses in Western countries. As has been observed by different authors and summarized, for example, in Fischer and Schot (1993), two phases of environmental responses of firms can be distinguished:

- from the 1970s till about 1985: a reactive approach and adjustment to regulatory pressure;
- from the mid-1980s: considerable change in firms' strategies and behaviour; institutionalization of environmental concern within firms shifting from a crisis-oriented, passive approach towards a cost-oriented (defensive) approach by the majority, and to innovative offensive approach by a few "environmental front-runners" (to use a phrase taken from Holm et al. 1994).

> The year 1985 is an arbitrary point of division. Nevertheless, somewhere in the mid-1980s firms changed from fighting or resistantly adopting to external pressures to embracing them and incorporating environmental considerations into their policies in a more rigorous way. (Fischer and Schot 1993:5)

This "arbitrary" year of 1985 correlates surprisingly well with the period of significant increase of environmental activities in Lithuanian companies. Why did it happen in the mid-1980s? Perhaps the internal changes in the USSR, the emergence of *perestroika* in 1986, played this decisive catalytic role, rather than "winds of environmental change" coming from the West. The gradual democratization slowly making its way in the Soviet Union, and more notably in its most "Westernized" part such as the Baltic states, fostered institutional changes and changes in the environmental authorities as well. The State Nature Protection Committee and its successor, the Department of Environmental Protection, developed out of changes in a number of ministries, boards and committees which were in charge of particular parts of fragmented environmental administration (e.g. the Board of Hydrology at the Ministry of Water Resources).

Although the phase from the mid-1980s is characterized in the West as a turning point towards proactive and innovative environmental management, it is difficult to characterize Lithuanian industry's environmental responses in the same way. The period from 1985 to 1989 was still a time of central planning. This means that the main motivations for being active in the environmental field were not market driven, i.e. it cannot be maintained that factors such as "competitive advantage" or "business image" were decisive. It is more likely that the main factors were the pressure from environmental authorities as well as the increased environmental awareness of particular industrialists (one might call them "champions" or "opinion leaders").

An important difference to the situation in the West was the ways in which funds were raised to implement waste and pollution reduction measures. There was little use of "Western" cost–benefit arguments. Instead, it was the same mechanism common to all activities in a centrally planned economy, namely "specific" entrepreneurship and bargaining skills of industrial leaders and their personal relations in ministries and committees in order to obtain limited funds or quotas. Because of this, particular companies (e.g. the slaughterhouses in Utena and Taurage) managed to purchase and install very modern, Western-made end-of-pipe control equipment or processing technologies that save significant amounts of water and energy (e.g. in the Utena dairy). The latter type of subsidized funding for technological change (e.g. the acquisition of a very modern process line in the wood chipboard manufacturer Giriu bizonas in Kazlu Ruda) was not specifically geared towards environmental protection; resource saving was just one positive "byproduct" of this process.

Most of this happened during several years in the late 1980s when, on the one hand, there still remained a system of planned economy, and hence opportunities to seek subsidies (in the ministries and committees in Moscow or Vilnius); and, on the other hand, companies got a chance to breathe some market economy air in terms of emerging possibilities to make direct contracts with foreign businesses. In most cases, companies could not use earned hard currency for paying wages. In consequence, they "had" to invest money in technological change. Some companies (e.g. the tannery Elnias in Siauliai) have built large local sewage-treatment installations. At present, however, due to the changed price structure for materials and operation costs, a number of companies have serious difficulties in utilizing end-of-pipe facilities.

Since the mid-1980s various companies (e.g. Kaunas's paper mill) have invested in local sewage-treatment facilities, but this construction was not finished by the time the planned-economy system of subsidies ceased to function. Such companies arrived at a complicated impasse: it costs a great deal to complete the construction which had already started, but at the same time companies have no funds to be able to operate even at a minimal level of production. Moreover, cost–benefit analysis does not prove that environmental technology solutions chosen because of availability of subsidies are most relevant in changed conditions.

Perhaps the most important bottleneck stemming from the Soviet era is a technocratic approach to industrial enterprise. It is not the machinery itself that is the problem; it is the lack of a long-term broad perspective, and the existence of a weak entrepreneurial "mentality" among Lithuanian industrialists which is a serious constraint on environmental innovation in companies. Some of the most important characteristics of environmental management currently promulgated in the discussions around the EU Eco-management and Audit Scheme, as well as ISO 14001, are information gathering, documentation, and exchange within the company, systematic assessment of problems and possible solutions, public disclosure etc. All these features require a different corporate management style and information culture, as well as communicative traditions which are not yet in place in many Lithuanian companies.

There are, however, some positive examples to point to, e.g. the former Kaunas Sweets factory which was privatized in 1993 by the Austrian company Kraft General Foods, a part of the Philip Morris group. This company manages continuously to screen waste and wastage reduction opportunities, assess potential measures and invest in new technologies at an amazingly high speed. However, it is hard to maintain that the management style of this company is based on a participatory approach. On the contrary, much of its success is a result of one strong leader (champion) who in pre-joint-venture times was production manager, chief technologist, chief energy manager and chief environmental manager all at the same time. At present this person (Mr Dziugys) controls the company's internal network and communication channels as well as channels of external techno-economic information. But this case of successful corporate environmental management is an exception for Lithuania, rather than the rule.

INCENTIVES AND CONSTRAINTS FOR WASTE AND POLLUTION MINIMIZATION IN LITHUANIAN INDUSTRY

To analize the variety of incentives and barriers for environmental management in Lithuanian industry in the period of transition, the framework has to accommodate concepts and issues comprising micro, meso, and macro level:

- micro level (companies' internal management structure, style, culture);
- meso level (relationships and networks in which the company is involved; connections with external actors which in different ways shape corporate willingness and ability to minimize waste and pollution);
- macro level (overall political, economic and social changes in the state that hinder or foster corporate greening).

Some incentives and barriers, for example, cover at least two levels of the analytical framework at the same time. The barrier of "too high interest rates to get a bank loan" (see Table 29.3) can be taken as an example. First, it has to do with macro-level trends in the country, for instance the high rate of inflation (it was about 40% in 1995). On the other hand, it reflects meso-level interorganizational issues such as a company's ability to demonstrate that it is a reliable business partner which will pay the loan back in time. A third component is interpersonal relations which are common for business relations in today's Lithuania (perhaps elsewhere as well). Often, it is not the company's business performance which guarantees a loan from the bank, but good personal relationships between two people or somebody's involvement in a particular network.

This is just one example of the complexity of issues and context in which companies are operating. For example, the issue of "debt for energy resources" (see Table 29.3), which is a heavy burden to many companies, is not such a narrow question as it might seem. It reflects the whole range of events and phenomena which occurred in Lithuania and other post-communist countries since 1990, and especially 1992. For instance, in 1992 the government of Lithuania issued a regulation which did not allow companies to use in Lithuania money earned by selling Lithuanian products in countries of the Commonwealth of Independent States (CIS). The companies were allowed to use this money only in the countries where these goods were sold, or to trade earned non-convertible currency with other companies which need it. In consequence, many companies could not recover their costs during half a year, a year or even longer, and their shortage of turnover cash was increasing dramatically.

In 1992, about 70% of the income of Lithuanian companies was from exports to the former USSR in unconvertible currency. Therefore, the vast majority of Lithuanian manufacturers had to delay significantly their payments of income and social insurance taxes and payments for energy resources. Simultaneously, the separation from the USSR led to a sharp increase in prices for energy and natural resources. For example, from 1991 to November, 1992, oil, gas and coal prices increased by about 15–20 times (Klevas and Tamonis 1992). In a sense, this shock for Lithuanian industry can be likened to the one that OECD countries experienced during the two energy crises in 1970s (Rinkevicius 1993).

As a result, many Lithuanian manufacturers have large debts to the state for energy resources and the size of this debt has grown dramatically over time. Therefore, it is unfair to look at the environmental management of Lithuanian companies from a perspective of a "typical" Western industrial enterprise. Very often, the company is willing to engage in substantial waste and pollution reduction, but has to cope first with problems on which its very existence and survival depend.

In addition, such important changes as privatization, shift in corporate ownership and management structure and style were expected to have important implications for corporate development in general, and environmental management in particular. For a few companies these changes had a stimulating effect (e.g. in the case of Kraft Jacobs Suchard Lietuva confectionery in Kaunas or Achema fertilizers' plant in Jonava). However, in many companies privatization caused a slow-down or even stagnation in production, sales and environmental activities.

The surveyed companies were asked to rank a list of incentives and barriers which foster or hinder waste and pollution reduction. The question on stimuli and answers to it (mean scores) are summarized in Figure 29.2 and Table 29.2; the question on constraints is reflected in Figure 29.3 and Table 9.3.

It is not surprising that the rapid increase in prices for energy and natural resources was given the highest score by Lithuanian companies as a stimulus for waste (and wastage) minimization. In order to control consistency of answers and have a deeper

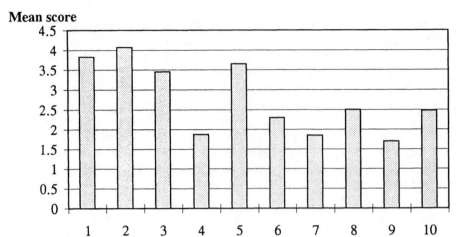

Figure 29.2 Incentives for waste and pollution minimization in Lithuanian industry

Table 29.2 Incentives for waste and pollution minimization in Lithuanian industry*

Incentive	Mean score	Standard deviation
1. Regulatory pressure, high pollution charges and fines	3.83	1.34
2. Rapid increase of prices for energy and raw materials	4.07	1.26
3. Expectation that in the future regulations will be more stringent	3.45	1.39
4. Public pressure (green movement, local communities)	1.87	1.02
5. Pursuit of minimization of negative impact on nature and humans	3.65	1.20
6. Environmental requirements imposed by owners and shareholders	2.30	1.29
7. Requirements of business partners (suppliers, customers, investors)	1.85	1.20
8. Attempt to catch up with competitors	2.50	1.41
9 Environmental requirements of Lithuanian banks and insurance companies	1.70	1.08
10 Ecological norms and standards to enter foreign markets	2.48	1.54

* Companies were asked to give to every incentive and barrier a score from 1 (unimportant) to 5 (very important). There were no restrictions on how many identical scores could be given to different variables.

Mean score

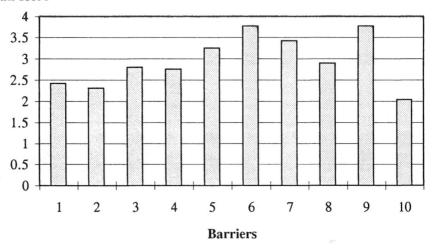

Figure 29.3 Barriers to waste and pollution minimization in Lithuanian industry

Table 29.3 Barriers to waste and pollution minimization in Lithuanian industry

Barrier	Mean score	Standard deviation
1. Lack of technical, economic and environmental information	2.42	1.24
2. Underdeveloped environmental legislation	2.31	1.28
3. Lack of economic interest among company employees	2.80	1.33
4. Lack of money due to significant debt to the state for energy resources	2.76	1.63
5. Lack of money due to significant debts between the company and its business partners	3.25	1.58
6. Underdeveloped general state tax system (VAT, import-export tax etc.) which does not promote investment in production	3.77	1.50
7. Problems in finding a new business niche	3.42	1.41
8. Problems in finding new stable suppliers of raw materials and related uncertainty about production and waste reduction	2.90	1.41
9. Too high interest rate to obtain loans from Lithuanian banks	3.77	1.55
10. Conservative, indifferent attitudes of new owners	2.04	1.38

insight into corporate motives, we included a question asking companies to rank the top three incentives and barriers. The ranking results show that, although growth of material/energy prices received the highest mean score (4.07), regulatory pressure was ranked as the first incentive most frequently (170 times), whereas growth of prices was ranked highest 137 times.

In general, both the field research and survey have shown that these two factors are undoubtedly the most important stimuli for waste minimization in Lithuanian industry. This corresponds with results of other surveys (e.g. by Steger 1993; Thidell and Arnfalk 1992) which found that regulatory pressure and economic benefits of environmental measures are the main motives fostering environmental management.

The third most important incentive found in our survey is willingness to minimize the negative impact on humans and environment (listed number 6 in the questionnaire). It is possible to wonder how objective such an answer is, because the survey may have had a certain bias. On the other hand, there is no reason to believe that Lithuanian industrialists are not conscientious in their answers or environmentally unaware, particularly those whose primary task in the company is to deal with environmental issues. Therefore, the ranking order seems justifiable.

The two most important barriers to waste and pollution minimization (both factors have the same mean score, 3.77) are located at the macro level reflecting overall state policy and processes of transition. These two constraints are the underdeveloped state taxation system (VAT, profit, import-export taxes etc.), and overly high interest rates for bank loans. Some aspects of the latter factor have been discussed earlier. The cumbersome and ineffective tax system obviously does not encourage investment in production and the development of environmental management systems. Of course, it would be unusual if business people and industrialists liked taxes. This topic, however, has specific characteristics in Lithuania and deserves special discussion.

Generally, in the ranking of the top three constraints, the taxation system was ranked highest most frequently (90 times) and as a second main constraint it was mentioned most frequently as well (77 times). We will return to the issue of taxes in the concluding remarks.

The third-ranked barrier is an inability or difficulty for many companies to find a new identity and new market niche in changed market conditions. This uncertainty of what to produce and where to sell a product makes it difficult for companies to define strategies for restructuring, and hence to improve environmental management. However, this problem reflects not only the internal characteristics of companies, e.g. lack of entrepreneurship, lack of skills in marketing and positioning etc. It reflects general "pains of transition" as well, because it deals with political, social and techno-economic changes in society.

For example, most furniture manufacturers had to reconsider their production and marketing strategy in 1995, because suddenly Russia and other CIS countries, the main market for many Lithuanian companies, introduced heavy import customs duties. In the case of furniture manufacturers, because of these duties, prices for some Lithuanian products increased by about 200%. As a consequence, these products could not compete on Russia's market with German or some other Western-made furniture, or even with that from Poland or Romania.

Thus the internal willingness of Lithuanian industries to change is very often inhibited by external constraints at a macrolevel. "But what about good housekeeping?", advocates of cleaner production would ask (Rodhe and Strahl 1995). Undoubtedly, incremental changes are very important. However, if one tries to go beyond simple clean-up of the shopfloor, any other changes (e.g. production process optimization, changes in operations and routines) need substantial knowledge, time and work. As one innovative industrial manager, Mr Abukauskas, director of the furniture company Freda, has put it, "Yes, good housekeeping is very important. But if I pay too much attention to how to protect my finger, I may forget about the risk of losing my head which is very likely in the current political and economic situation."

There are very few companies in Lithuania which already think along the lines of ISO 14001 or EMAS. Most companies are familiar with waste and pollution reduction, some of them with pollution prevention and cleaner production (some are doing waste minimization without calling it that). However, the number of those which seek to establish environmental management systems is very small. Such companies are usually either totally dependent on the foreign market (e.g. the fertilizers plant Achema in Jonava) or are subsidiaries of innovative Western companies (e.g. Novotex from Denmark, the leader in "green cotton" business).

CONCLUSIONS

Environmental policy changes are not likely to achieve their ultimate goals if they are not correlated and integrated with changes in other governmental policy domains. First of all, the overall taxation system, and the environmental charge system in particular, has to be revised. This work is already underway in Lithuania. It is hoped that the working group on environmental charge reform established at the Ministry of Environmental Protection will be able to prepare consensus-based policy documents which would be acceptable to both the government and industrialists.

An establishment of alternative financial institutions and schemes is urgently needed to support waste and pollution minimization in industry. The existing system of commercial banks is not prepared to provide such support to a large number of companies. One model of alternative financing is the establishment of a revolving environmental fund. Similar funds have already been set up in Poland, Hungary, the Czech Republic and other CEE countries. On the one hand, such a fund would provide soft (subsidized) loans to industry, and on the other hand, it would accumulate part of the environmental assistance money. Foreign donors who want to support the greening of Lithuanian industry in this case would deal with one institution set up in cooperation by environmental authorities, industrial organizations and other stakeholders. Establishment of such a fund is speeding up in Lithuania with assistance and support from USAID and the EU Phare programme.

It is worth considering the possibility of channelling to the revolving fund at least part of the money collected as environmental charges and fines which at present goes to municipal and central republican funds (in proportions of 70% and 30% respectively). In this way, the overall institutional arrangement would allow companies to recover part of the money paid as environmental charges in the cases when companies themselves propose viable waste and pollution minimization plans.

Information, education and retraining of industrialists is another policy domain which needs to be developed. The projects on waste minimization and cleaner production that are already conducted or initiated in Lithuania by US, Scandinavian, Dutch and other national and international organizations provide a very good starting point. "Train the trainer" is probably the most effective approach to disseminating education and information on pollution prevention and environmental management.

The question arises of whether all educational activities and assistance projects should be coordinated and carried out by one national centre, i.e. in a fashion similar to central planning. This approach appears in working papers drafted for a meeting at OECD, Paris, held in June 1995, in a series of preparatory meetings for the Sofia ministerial conference in October 1995 (within the framework of the Eastern European Environmental Action programme). However, the recommendation to support just one "national" pollution prevention or cleaner production centre might create an unhealthy atmosphere and induce competition in the country. One can already notice symptoms of this in relations between stakeholders in Lithuania and Poland.

It might be suggested that information and education should be provided by a number of centres, institutes and other organizations. This approach would allow better coverage of various geographic areas and access to specialized knowledge and expertise available from various organizations and individuals. In this way, the system would be more like a network of travel agencies each having specific advantages, while all serving the same purpose, rather than a system such as Intourist in the former USSR which, due to its monopolist position gradually became inflexible and could not provide quality services to customers. Of course, coordination is necessary between various organizations promoting pollution prevention and environmental management. But probably it is better that such a network develops from the bottom up, following the model, for example, of the US National Pollution Prevention Roundtable.

REFERENCES

Brundenius, C. (1995) "How painful is the transition?" in Lundahl, M. and Ndulu, B.J. (eds) *New Directions in Development Economics? Growth, Environmental Concerns and Government in the 1990s*, London, Routledge.

Fisher, K. and Schot, J. (1993) "The greening of the industrial firm", in Fisher, K. and Schot, J. (eds) *Environmental Strategies for Industry*, Washington, DC, Island Press.

Freeman, C. (1991) *Technology and Future of Europe: Global Competition and Environment in the 1990s*, London, Pinter Publishers.

Freeman, C. and Perez, C. (1988) "Structural crises of adjustment, business cycles and investment behaviour", in Dosi, G., Freeman, C., Nelson, R., Silverberg, G. and Soete, L. (eds) *Technical Change and Economic Theory*, London, Pinter Publishers.

Halsnaes, K. and Morthorst, P.E. (1990) *Sustainable Development as a Planning Goal for Energy Systems in the Nordic and Baltic Sea*, Denmark, RISO, J.1108 90–27.

Hoffman, A. (1995) "The environmental transformation of american industry: an institutional account of organizational evolution in the chemical and petroleum industries (1960–1993)", PhD dissertation, Cambridge, Massachusetts Institute of Technology.

Holm, J., Klemmensen, B. and Stauning, I. (1994) *Two Cases of Environmental Front Runners in Relation to Regulation, Market and Innovation Network*, Department of Environment, Technology and Social Studies, Roskilde University.

Jamison, A., Rinkevicius, L. and Strahl, J. (1994) *Cleaner Production: A Cognitive Approach*, paper presented at the Third Conference of the Greening of Industry Network, "From Greening to Sustaining: Transformational Challenges for the Firm", Copenhagen, November.

Jankauskas, V. (1992) "Improving energy efficiency in Lithuanian industry: difficulties of transition", in *Improved Energy Efficiency in Former Centrally Planned Economies*, proceedings of the IAEE East European Conference, Kaunas, Lithuania.

Klevas, V. and Tamonis, M. (1992) *The National Program of Energy Saving*, Vilnius, Ministry of Energy of the Republic of Lithuania.

Mol, A.P.J. (1995) *The Refinement of Production: Ecological modernization Theory and the Chemical Industry*, Utrecht, Van Arkel.

Nelson, R.R. and Winter, S.G. (1977) "In search of a useful theory of innovation", *Research Policy*, Vol. 6, pp. 36–76.

Rinkevicius, L. (1992) *Cleaner Technology in the Transition from Centrally Planned to a Market Economy: the Case of Lithuania*, paper presented at 4S/EASST Joint Conference, "Science, Technology and 'Development' ". Gothenburg, August 12–15.

Rinkevicius, L. (1993) " 'The double transition': prospects of cleaner production in Lithuania", Master's thesis, Lund, Research Policy Institute, Lund University.

Rinkevicius, L. (1994) " 'Aplinka, technologija ir visuomene" ("Environment, technology and society"), *Philosophy, Sociology*, Nr. 3, Vilnius, Lithuanian Academy of Sciences.

Rodhe, H. and Strahl, J. (1995) "Western support for cleaner production in Central and Eastern European industry", *Business Strategy and the Environment*, Vol. 4, Nr. 4, pp. 173–179.

Steger, U. (1993) "The Greening of the board room: how German companies are dealing with environmental issues". In Fisher, K. and Schot, J. (eds) *Environmental Strategies for Industry*, Washington, DC: Island Press.

Thidell, A. and Arnfalk, P. (1992) *Environmental Management in the Swedish Manufacturing Industry*, Lund, Department of Industrial Environmental Economics, Lund University.

Wallner H.P. and Fresner, J. (1995) *Industrial Networks at the Regional Level—Creating Islands of Sustainability—Theory and Practice*, paper presented at the Fourth Conference of the Greening of Industry Network, Toronto, November 12–14.

30
Implications of Czech cleaner production case studies for environmental management systems

VLADIMÍR DOBEŠ

INTRODUCTION

The United Nations Industrial Development Organization (UNIDO) has carried out a survey of developing countries and economies in transition to identify the specific problems related to introduction of ISO 9000 and ISO 14000 series (UNIDO, 1995). Of the factors that were cited by respondents to the survey questionnaire as deterring companies from implementing ISO 14001, the most common was a lack of awareness of the benefits. The survey showed that the main reason for implementing ISO 14001 was just to demonstrate conformity to legislation.

There is a real danger that industrial enterprises perceive environmental management systems (EMS) as a certificate which needs to be obtained because otherwise it acts as a new trade barrier. The purpose of this chapter is to demonstrate the possible benefits of the introduction of an EMS by building on the results of cleaner production (CP) case studies implemented in Czech industrial enterprises and to discuss possible implications of CP for EMS.

Environmental Management Systems and Cleaner Production, edited by R. Hillary.
© 1997 John Wiley & Sons Ltd.

BACKGROUND INFORMATION

Czech industry is now facing two imperatives: one to increase its productivity and competitiveness; and the other to improve its environmental performance. These two imperatives used to be perceived as incompatible; however, the situation is changing and more and more enterprises in the Czech Republic are undertaking voluntary activities to improve their environmental performance and increase competitiveness at the same time.

CP brings the prevention of wastes at source, the more efficient use of raw materials and energy and the potential to save costs relative to end-of-pipe treatment (for more information see UNEP 1994). The number of CP projects in Czech industrial enterprises is increasing: in 1992 there were 11 projects, 1993 10 projects, 1994 13 projects and 1995 25 projects.

EMS are quickly becoming an issue of interest mainly for big Czech enterprises exporting abroad. Three such enterprises started preparation on certification of EMS in 1995 and many other enterprises are investigating opportunities to start implementing EMS. Those which already have experience with standards in the ISO 9000 series are interested in ISO 14001.

CLEANER PRODUCTION CASE STUDIES

Czech–Norwegian Cleaner Production Project

The case studies, which are analysed below, were implemented within the frame of the Czech–Norwegian Cleaner Production Project in the period 1992–95. The development objective of the project was to create a domestic capacity in CP. Within the framework of this project the first 122 Czech professionals were trained, the first 34 case studies developed and the Czech Cleaner Production Centre (CPC) was established.

A "train the trainer" approach was used by the Norwegian Society of Chartered Engineers (NSCE) and implemented in the form of long-term interactive training programmes lasting eight months. The programmes included theoretical lectures and exercises and on-the-job training using concrete case studies in industrial enterprises (for more information see Nedenes 1995).

Method

The method used to implement the case studies included the following main steps:

- getting top management commitment;
- planning and organization (steering and working group);
- preparation and publication of enterprise environmental policy;
- pre-assessment and selection of topic for case study;

- assessment—identification of causes of waste generation (e.g. by implementing material balances);
- generation of CP options;
- feasibility study;
- implementation of CP measures;
- measuring benefits of CP measures and evaluating results;
- keeping the CP measures ongoing;
- follow-up planning (CP programme).

The implementation of the above method in enterprises is referreed to as CP case study or CP project in this chapter (for more information see US EPA 1992).

CP projects are directed at identifying gaps in the enterprises management system by solving concrete technological problems. While the EMS is a top-down approach, CP is more of a bottom-up approach. There is a similarity between CP and EMS in the sense that both require effort directed towards continual improvement.

The following results from CP case studies demonstrate the possible benefits of EMS when the preventive strategy is used systematically.

Results of case studies

This overview of results (evaluated on site by CPC (1995) includes only those CP measures which have been implemented or which are to be implemented (as decided by company management) in the very near future. For this reason 10 of the 34 implemention case studies have been excluded. This is stressed to illustrate that the presented results are practical not theoretical. The case studies' credibility is reinforced by the fact that costs for external assistance to the 10 case studies implemented during 1994/95 were partly paid by the enterprises involved from proven benefits of CP measures.

Summary of the main quantifiable benefits from 24 CP case studies

1. Total economic benefits:
   - more than CZK80 million per year
2. Total environmental benefits:
   - 2100 t/y of VOC
   - 12 000 m^3/y waste water
   - 21 500 t/y of industrial waste (from this 9500 t/y is hazardous waste)

These cummulative results from the 24 case studies are presented in Figure 30.1. Results from 10 case studies implemented in 1994–95 are presented in more detail in Figure 30.2.

The savings from not investing in planned end-of-pipe technologies have been omitted from the project evaluation and are not presented in the figures. This was investigated during evaluation of the programmes during 1994–95 and the expert

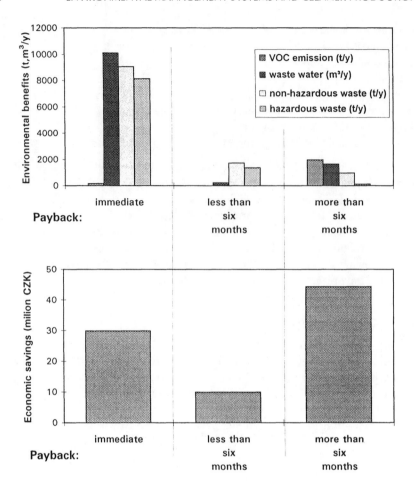

Figure 30.1 Cumulative results from 24 case studies implemented in 1992–95

estimate showed that the economic savings by reducing or avoiding investments in end-of-pipe technology are more than CZK30 million.

End-of-pipe technological solutions and command-and-control regulations will maintain thier importance into the near future. However, CP project implementation before an application of end-of-pipe technology has the benefit that the end-of-pipe solutions can be reduced or even avoided by CP implementation.

For example, the KOH-I-NOOR Praha enterprise entered the above described CP course because of the high costs of a planned waste water treatment plant (US$792 000)—summary results are presented in Figure 30.2. By the end of the case study, the investment costs for the neutralization station were US$538 000.

not include time spent by course participants at the training centre—15 days per participant). Figure 30.2 shows that the better results from case studies correspond to the higher labour inputs. The minimum labour input for a good case study could be estimated to be between 30 and 40 persondays. Integration of CP into EMS could result in an increase in labour input because of the system operations necessary when implementing CP projects instead of routine application of end-of-pipe solutions. There is no data available to draw a conclusion but the CP case studies show that if the EMS works with CP payback, costs for the EMS, including its certification, could be recouped.

Integrated management

CP can improve the competitiveness of enterprises. Improved competitiveness comes not only from environmental benefits but also product quality, the working environment and corporate culture and image. The variability of CP effects in different directions is illustrated in Figure 30.2. The case studies in the figure illustrate a mutual interconnection between these areas and the importance of their balanced integration into the enterprise's management system (Zwetsloot 1994).

CONCLUSION

The final aim of the projects described was not the cases themselves but the building of interim professional management capacities and the demonstration of the benefits of CP to start an enterprise CP program. This final aim was met only in enterprises where CP was integrated into their management systems. Therefore, we perceive EMS as an important tool to make the process of CP implementation sustainable.

In conclusion, CP projects are a good initial step to establishing EMS and promoting CP, as the aim of EMS is to realize the win–win effect for all stakeholders interested in continual improvement of environmental performance in industry.

REFERENCES

Czech Cleaner Production Centre (1995) *Czech/Norwegian Cleaner Production Project 1992–1995—Final Report*, Prague, CPC.

Nedenes, O. (1995) *Best Practice Guide for Cleaner Production in Central and Eastern Europe*, Paris, OECD.

United Nations Environment Programme (UNEP) (1994) *Industry and Environment*, Vol. 17, No. 4, Oct–Dec, Paris, UNEP.

United Nations Industrial Development Organization (UNIDO) (1995) *Trade Implications of International Standards for Quality and Environmental Management Systems (ISO 9000/ISO 14000 Series)*, Draft Report, Vienns, UNIDO.

US Environmental Protection Agency (EPA) (1992) *Facility Pollution Prevention Guide*, Washington, US EPA.

Zwetsloot, G. (1994) *Joint Management of Working Conditions, Environment and Quality*, Amsterdam, NIA.

TON as BYSTŘICE pod HOSTÝNEM

Industry: Furniture	Subject of study: Minimization wastes from surface finishing of furniture	Turnover(milion)/No of employees: 371/1100

Total effects	Economic savings	Payback	Waste reduction	Main reduced waste flow
	4767 thousand CZK/y	Immediate	47.4 t	Emission of solvents

Graph values: WR = 25.6% emission of solvents, SP = 4 668 000 CZK/y, SW = 294 000 CZK/y. The saving of 138 t/y (25.6%) of synthetic coatings is the main benefit reached by a simple change of the existing technology. The operation costs were increased by 375 000 CZK/y at the same time. Total labour intensity is 150 persondays.

VUAB sp Roztoky u Prahy

Industry: Pharmaceutical	Subject of study: Minimization of waste from ephedrine production	Turnover(milion)/No of employees: 600/550

Total effects	Economic savings	Payback	Waste reduction	Main reduced waste flow
	6022 thousand CZK/y	Immediate	73 t	Butyl-acetate

Graph values: WR = 18.6% waste butyl-acetate (from this 52 t air emmission and 21 t waste water). SP = 2 539 000 CZK/y, SW = 2 483 000 CZK/y Another benefit is diminishing of COD by 526 000 kg O₂. Improvement of WE is given by reduction of butyl-acetate emission. Total labour intensity is 250 persondays.

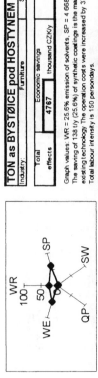

Figure 30.2 (continued)

MORAVOLEN SUMPERK as

Industry:	Textile	Subject of study:		Turnover(million)/No of employees	771/1940
Total effects	Economic savings 14480 thousand CZK/y	Payback 4.1 years	Waste reduction 784 t	Main reduced waste flow Utilisation of flax waste Flax waste (dust and fibres)	

Graph values: WR = 100% flax wastes, SP = 14 480 000 CZK/y
Modernization of technology enables more effective use of raw material and saves 172.5 t/y of flax waste.
The remaining fibre waste will be recycled onsite and the dust waste will be processed to a marketable product.
Total labour intensity is 120 persondays.

NASIN sro Liberec

Industry:	Textile	Subject of study:		Turnover(million)/No of employees	40/120
Total effects	Economic savings 361 thousand CZK/y	Payback 0.5 year	Waste reduction 227 m3	Main reduced waste flow Minimization of waste size Waste size	

Graph values: WR = 91.5% waste size (from this 85% reached by prevention of waste dilution), SP = 14 000 CZK/y, SW = 347 000 CZK/y. The savings of 25 000 CZK and reduction of waste flow by 6.5% (estimation) were achieved by good housekeeping measures.
Total labour intensity is 15 persondays.

TECHNOPLAST as CHROPYNI

Industry:	Plastics materials	Subject of study:		Turnover(million)/No of employees	1500/1700
Total effects	Economic savings 4767 thousand CZK/y	Payback 2 years	Waste reduction 62 t	Main reduced waste flow Minimization of wastes from production of syntetic leather Emission of solvent (DMF)	

Graph values: WR = 44.4% emission of dimethylformamide, SP = 4 663 000 CZK/y, SW = 104 000 CZK/y
The energy savings of 2 810 000 CZK and 1561 m3 of water are included in presented results.
Improvement of WE is given by diminishing the DMF emission into the working environment.
Total labour intensity is 100 persondays.

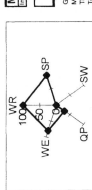

Figure 30.2 (continued)

HRUSOVSKA CHEMICKA SPOLEENOST sro

Industry:		Subject of study:		Turnover(milion)/No of employees	524/500
Chemical				Minimization of baryte sludge	
Total effects	Economic savings	Payback	Waste reduction	Main reduced waste flow	
	3336 thousand CZK/y	0.5 year	1310 t	Baryte sludge	

Graph values: WR = 22% waste sludge, SP = 2 863 000 CZK/y SW = 472 000 CZK/y Another contribution is better utilization of raw materials-baryte and coke-and from this resulted savings of 9311t/y of baryte and 379 t/y of coke. The presented data are based only on laboratory experiments for the time being. Total labour intensity is 25 persondays.

KOH-I-NOOR PRAHA as

Industry:		Subject of study:		Turnover(milion)/No of employees	180/433
Metal-finishing				Minimization of waste from galvanizing shop	
Total effects	Economic savings	Payback	Waste reduction	Main reduced waste flow	
	907 thousand CZK/y	immediate	8250 m³	Waste water	

Graph values: WR = 32% waste water, SP = 802 000 CZK/y, SW = 105 000 CZK/y
154 000 CZK/y from this was saved by investment with a payback 0.5 year. Other benefits are energy savings and reduction of used chemicals by 3 t/y.
Total labour intensity is 40 persondays.

MIKOV MIKULASOVICE s.p.

Industry:		Subject of study:		Turnover(milion)/No of employees	110/340
Metal-finishing				Minimization waste from galvanizing	
Total effects	Economic savings	Payback	Waste reduction	Main reduced waste flow	
	172 thousand CZK/y	0.5 year	6950 m³	Waste water	

Graph values: WR = 37% waste water, SP = 116 000 CZK/y, SW = 56 000 CZK/y
Another benefit is reduction of waste sludge by 10 t/y (12.5% from the total production).
The existing electrodialysis unit will be used more effectively by implementing simple CP measures.
Total labour intensity is 40 persondays.

Figure 30.2 (continued)

Description of axis:

WR-waste reduction (% reduction of the main waste flow)

Note: US$ 1 = ca CZK 26

SP-savings on inputs and process (thousand CZK/year-e.g. savings on raw materials and energy, dimension of axis is 2% of turnover)

SW-savings on waste treatment (thousand CZK/year-e.g. savings on transport and waste disposal, emission charges, etc., dimension of axis as SP)

QP-increase of production quality (%-e.g. reduction of aborted products or increase in product quality)

WE-improvement of workplace environment (%-e.g. dust reduction or emission reduction at the workshop)

The figure indicates CP benefits in different directions. It is not possible to compare values on axis SP and SW at different enterprises because of differences in size of enterprises and problem investigated.
SP and SW represents ratio of economic savings to enterprise turnover related to 2% of turnover. WE and QP are an expert estimation. Only in the case of KOH-I-NOOR (60%) and MIKOV (50%) are they given by measurement of rejected parts before and after implementation of CP measures. The data on labour intensity are only raw estimates and do not include time on theoretical training.

Figure 30.2 Results from 10 CP case studies implemented in 1994–5

KOH-I-NOOR Praha continued with the CP programme without external assistance and reached further reduction in the required capacity of the waste water treatment plant end-of-pipe technology. The required investment is currently only US$300 000 and reduction in operating costs is estimated to be approximately US$58 000 per year.

IMPLICATIONS FOR ENVIRONMENTAL MANAGEMENT SYSTEMS

The following analysis is of the presented results from the point of view of possible implications for EMS.

Potential for good housekeeping

Significant potential for economic savings and environmental improvements exists in Czech industry and can be exploited through CP. A large part of these benefits can be achieved by simple good housekeeping measures. The potential benefits of EMS systematically using the CP method are those derived from the organizational potential for pollution prevention. The average annual savings per company from implementing good housekeeping measures (most of which were implemented during the CP training programme) were more than CZK1 million (approximately US$50 000), around 10 times more than the costs of the external assistance to implement the case study i.e. the training course and the methodological guidance of the case study, in each enterprise, see Figure 30.1. The conclusion from this is that, by using CP, enterprises can meet the aims of their environmental management policy in an effective way.

Information system

The major contributor to economic savings was increased efficiency in the production process. Savings on raw materials, energy, work etc. were, on average, 10 times higher than savings on waste treatment. This was investigated in detail during the course of 1994–95 and is illustrated in Figure 30.2 (see axes "savings on process" and "savings on waste"). The savings on the minimization of waste flow itself (end-of-pipe treatment) were only marginal (10% in average). This saving often surprised the enterprises. The information system within EMS should provide the necessary information on material and financial flows to point out the potential for CP.

Compensation of labour intensity

The average total labour input of each case study (in–company work) was around 25 persondays per enterprise, although this varied from 15 to 250 persondays (this does

31

The combined introduction of environmental management systems and cleaner production in industry

GULBRAND WANGEN

INTRODUCTION

The subject of this chapter is cleaner production (CP) programmes and environmental management systems (EMS), both of which are tools to improve environmental performance in business and industry. CP programmes are used primarily in industry to introduce clean technology and enhance capacity building for application of cleaner production assessments. EMS are being developed for, and introduced, in a range of sectors and the focus is on management practices. The two different approaches are discussed separately in this chapter and the situations in Norway and in Hungary in terms of legislative requirements and the needs and challenges of industry are considered. The use of these tools in a project in Hungary are highlighted.

Environmental Management Systems and Cleaner Production, edited by R. Hillary.
© 1997 John Wiley & Sons Ltd.

TODAY'S TRENDS

International trade and harmonization

The World Trade Organization (WTO) and regional treaties such as that for the single European market and NAFTA open up markets and liberalize trade. In order for this to work without distorting competition, a "level playing field" has to be created. Standards are thus playing an increasingly important role in conformity assessment globally. The International Organization for Standardization (ISO) is playing a significant part in this process. The new role of standards has also led to a much more active involvement in standard making, which in principle is open to all potential users.

New, harmonized and stricter legislation

Legislation has become stricter and it will no doubt continue to do so. Internationally, regulations on strict, no-fault environmental liability are also being introduced or are in the pipeline. These regulations will have major consequences for business.

Furthermore, legislation is increasingly becoming function oriented as opposed to prescriptive. It is more natural and resource efficient to make companies take responsibility for environmental performance rather than prescribing detailed equipment and operating requirements. Norway was probably the first country to pass legislation requiring companies to set up an integrated management system covering health, safety and environmental (HSE) protection based on quality management thinking.

Legislation is also being harmonized. For instance, the Maastricht treaty of the European Union (EU) introduces majority voting in the Council of Ministers for environmental issues to avoid poor environmental performance becoming a competitive advantage for less developed countries.

In WTO, the question of social issues (children's work, health regulations, working hours etc.) has become very controversial. Environmental issues will become the next issue for nations to fight over. This raises the issue of "green protectionism", i.e. the allegation that rich countries are using high environmental standards and their technical lead to secure their competitive position.

One example is Norway's interest in taking a lead in ISO/TC 207's subgroup on environmental performance evaluations. When developing performance indicators for industry production Norway wants its vulnerable recipients (unpolluted fjords) and its clean energy (hydropower) to be taken into consideration in rating the environmental impacts of, say, aluminium production. By using a life cycle assessment (LCA) approach to the emission of greenhouse gases and environmentally damaging discharges, the resultant pollution is less than that produced from the same amounts of metal manufactured using coal in densely populated Central Europe.

Whatever the motives behind the initiatives, the trend points to environmental performance becoming an important competitive factor in the future.

ENVIRONMENTAL MANAGEMENT SYSTEMS (EMS)

EMS is a fairly new concept. The British Standards Institution (BSI) introduced the world's first EMS standard in 1992. The BSI (and also to some extent British industry) experienced success (and a competitive advantage) when ISO adopted the BSI quality standard (BS 5750) as the ISO quality standard (ISO 9000 series). EMS is now subject to standardization in ISO in a Technical Committee which consists of the following subgroups:

- Environmental Management Systems
- Environmental Auditing
- Life Cycle Assessment
- Environmental Performance Evaluations
- Environmental Labelling
- Terms and Definitions

International standardization work is progressing rapidly and it is clear that most countries are trying to influence the developing standards to suit their home users.

At the European Union level the Eco-management and Audit Scheme (EMAS) has been operational since April 1995. EMAS is a voluntary regulation where participating companies must establish and implement an EMS, perform systematic, periodic and objective evaluation (environmental auditing) of the performance of such systems and provide information to the public on environmental performance.

From a business perspective, the motivations for establishing an EMS are:

- better management control of environmental performance;
- a system to ensure compliance with legislative requirements based on proven management principles;
- and the fact that EMS may be required by the market or can be of use in marketing.

CLEANER PRODUCTION (CP)

In 1988, the US Environmental Protection Agency (EPA) published a manual on *Waste Minimization and Opportunity Assessment in Industry*. The intention was to identify improvements in environmental performance that were economically advantageous. The manual was amended in 1992 and renamed *Facility Pollution Prevention Guide*.

This manual introduces simple methodologies for calculating mass and energy balances for parts of production where environmental problems are known to occur or for identifying environmental problems. Based on the balances, options for rectifying identified problems are developed and by way of simple economic analysis payback periods or other economic indicators are used to implement so called win–win strategies. The methodology has been adopted internationally and the EPA manual has been translated and modified in many countries. In Norway, translation was completed in 1991.

SITUATION IN NORWAY

Industry and the authorities

Norway is characterized by being:

- not heavily industrialized;
- advanced in integrated pollution control.

Norway has relatively few industries compared with other European countries. The industrial structure consists of small and medium-sized enterprises in combination with several heavy industries located at strategic sites. Heavy industry is primarily energy intensive, typically with a low degree of value adding, i.e. there is little production of consumer goods. (Consequently it was hard for other countries to find Norwegian consumer goods to boycott when Norwegian whaling was under attack.) Norway's industrial profile has led to few traditional environmental problems, although locally the metallurgical industry, mines etc. have created environmental problems both in terms of pollution and ground contamination.

Norway was the first country to establish a Ministry of the Environment, in the early 1970s. Its environmental legislation is considered advanced and integrated pollution control based on an individual licensing system which takes account of recipient conditions has been in place for many years. Industry, however, has increasingly complained about the growing number of regulations in the areas of health, safety and environment (HSE), as well as in other areas.

The management system approach

In the mid-1980s, the authorities took a critical look at their achievements in the area of HSE. Although Norway was still a good performer in this area by international standards, it was concluded that the results were not commensurate with the ambitions and resources used. This conclusion was a "moment of truth" and a reorientation was called for in the way in which the authorities were working with companies.

The reorientation happened in the mid-1980s at around the same time as the introduction and the rapid support of standards in the ISO 9000 series on the quality management. ISO 9000 sets out principles for the establishment of procedures to ensure the quality of products. Implicit in the standard is the concept of continual improvement of a company's systems, i.e. policies, goal-setting procedures, instructions and organization of everything around the value-adding process.

Norway thus became one of the first countries to introduce function-oriented legislation in the areas of HSE. How the company was to comply was to be determined by the company. Under the internal control (IC) regime companies had to set up management systems to assure the authorities that they had systems capable of ensuring compliance with all applicable laws and regulations. A systematic approach is required on the part of the company, including:

- analysing all important processes;
- identifying applicable requirements, in this case legislation;
- developing effective control mechanisms;
- documenting control mechanisms.

The IC regime parallels the implementation of a quality system where the resulting documentation, among other things, is a quality manual to be used by staff and shown to customers. The ISO 9000 idea is to show customers how all processes are carried out under controlled conditions to ensure the acceptable quality of final products.

In an IC system the resulting documentation is an IC manual showing how the company adheres to legislative requirements. This manual is the starting point of a system audit by the authorities. Such a system audit will be parallel to a quality system—or an EMS audit—and will typically conform to the following format:

- Review of the manual to see that its descriptions cover the company's processes and applicable laws.
- Audit of the site to see that the actual practices correspond with the descriptions in the manual.
- Verification by measurement and traditional inspection activities.
- Closing meeting with the company's management to present the impressions of the quality of the system, any deficiencies, and recommendations and instructions to rectify the system with the IC regime in order to comply.
- Reporting to management the audit results.

Both industry and the authorities intend the IC system to improve performance and reduce the need for detailed inspections. Furthermore, it is interesting to note that deficiencies in the company's management system are a breach of the law, even though all emissions, discharges etc. are otherwise satisfactory.

The Norwegian Cleaner Production Programme

In parallel to the IC approach, CP was promoted by the Norwegian Pollution Control Authority (NPCA) in a widespread programme initiated in 1990. The programme, adopted from the US EPA, had the following objectives:

- to shift the focus to preventive environmental management;
- development of cleaner technologies in Norway;
- to stimulate companies' interest in carrying out analysis;
- to build up competence among consultants;

- to achieve national goals by reducing emissions, discharges and waste generation efficiently, i.e. through "win–win" strategies;
- to work "with" the industry rather than "against" it.

Financial support from the NPCA was given to two types of projects:

- demonstration projects aimed at developing or introducing new technology for improving the environmental situation;
- CP assessments for identifying improvements.

An independent organization, the Norwegian Society for Chartered Engineers (NIF), was made responsible for the promotion and implementation of the capacity-building programme in industry, because industry has more faith in NIF than in the authorities and because NIF would be in a better position to motivate companies to participate.

The financial package prepared for implementing CP assessments in companies was as follows:

- Small companies: 85 consulting hours; medium-sized companies: 110 consulting hours; large companies: 135 consulting hours paid for by NPCA.
- Companies match and pay for the same number of consultant hours as well as providing their own manpower resources which are greater than the consultant assistance.

The cost of the capacity-building programme so far has been US$215 000 for information/dissemination and US$1.3 million for grants to do assessments. Assessments were completed in 246 companies by the end of 1994.

The results of the programme are summarized as follows:

- implementation of cost-effective "housekeeping" actions and minor process changes;
- enthusiasm among employees for "housekeeping" measures;
- reduction of wastes and discharges by 20–50% as well as the segregation and external recycling of wastes and improved pollution control.

As with many other CP programmes, the Norwegian experience shows that top management involvement is crucial for its success. Top management commitment ensures that the company sets aside enough time and allocates enough resources for the implementation of a CP assessment. The choice of consultant is also very important: an energetic consultant can have a significant impact on the final results. Finally, it is important to clarify the role and the division of responsibility between the company and the consultant. The rule is that the CP assessment should not be just another consultant report but should be based in an internally driven process supported by company top management.

The reasons some companies are not interested in participating in CP projects is that they have limited resources (especially of key personnel), have many other offers and/or have difficulties in relating the CP assessments to other management challenges in the area of HSE.

The Norwegian CP programme of grants to carry out CP assessments in industry is due to be reduced and more emphasis will be put on information about the potential of CP in selected industrial sectors. Also the licence application procedure of the Pollution Control Act will be extended to include a requirement for CP assessments when appropriate.

SITUATION IN HUNGARY

Environmental legislation and responsibilities

The rapid transformation of Hungary since 1989 when it introduced a market-based economy has presented new challenges and external requirements for industry. Hungary's Environment Protection Act II (1976) forms the basis of its pollution-control measures and the related application and permit procedures. The 1992 privatization legislation included additional environmental damage remedies in the privatization process and requirements for informing the environmental authorities about planned remedies.

Most public environmental protection activities in Hungary are under the jurisdiction of the Minister of the Environment. The Minister supervises a national service called the National Environment Inspectorate. There are 11 regional inspectorates serving as independent units (first-instance authority). The regional inspectorates implement and enforce regulations. In addition to these inspectorates, the National Public Health Service, under the Minister of Public Welfare, has powers in matters of air quality.

National environmental protection policies are determined primarily by regulations enacted by the Ministers of the Environment and Public Welfare. Environmental permit requirements and industrial and building permits are generally media specific.

The government plans to privatize up to 60% of state-owned assets by 1994. In the various privatization deals involving foreign investors, liability for past environmental damages often causes delays and uncertainties.

When privatizing and transferring the ownership of companies from the state to the private sector, investigations will have to focus on the environmental risks associated with a company and site. Environmental permits are not reviewed during the process of privatization or transfer. There is no requirement to prove ability to comply with the permits.

In the privatization process the following regulatory issues are typically investigated:

- What permits or approvals are required to operate the property to be transferred?
- Does the present facility have all the necessary permits and approvals?

- Are these permits and approvals valid?
- Is the present facility in compliance with the terms and conditions of its permits and approvals, or other hazardous substance requirements, or environmental statutory or regulatory requirements?
- Can the site or facility, as presently configured, comply?
- What permit transfer difficulties may be encountered? For example, are the permits transferable? Must new permits be obtained? Are public hearings required on permit transfer or the issue of new permits?
- Regardless of regulatory compliance issues, are there other civil liabilities associated with the site or facility stemming from hazardous substance or other environmental conditions?

Non-compliance with legislation or permit requirements has a number of administrative consequences. Administrative measures range from the determination of specific duties of operation and setting special conditions of operation to the temporary limitation of operations. Pollution fines are levied for excess of emission standards or unlicensed production of waste; however, major air-polluting industries pay reduced fines due to regulatory exemption.

New plants are generally subject to more stringent environmental requirements than old plants in the sense that they must meet certain national and local requirements during the planning stage.

Based on a hypothesis that the privatization process should open up the need for improving environmental management systems and implementing CP assessment in industry to identify cost-effective investments and related environmental performance improvements, DNV, together with its partners, has tried to adapt an integrated quality and environmental management system (QEMS) in a Hungarian industry.

QEMS IN HUNGARY

At the end of 1992, DNV started cooperating with Dunaferr Share Company with the aim of developing a management system covering quality and environmental matters for one division of the company. The system was to be based on the two internationally established standards, ISO 9000 on quality and BS 7750 on the environment. The system was called quality and environmental management system or QEMS.

The project was financed by the Norwegian Ministry of the Environment under its special grant programme to Central and Eastern Europe.

The objectives of the project were:

- development and implementation of a QEMS system suitable for the company;
- knowledge transfer and training in quality and environmental management thinking and methods of CP;

- Tangible improvements in quality and the environmental situation;
- dissemination of results to Hungarian industry.

The company

The project participants included a large coke plant where the QEMS was to be implemented and an in-house consultancy delivering quality and environmental services to the remainder of the organization. The results were disseminated to the rest of the steelworks through a top management seminar.

The Dunaferr steelworks is the largest steel manufacturer in Hungary and is a state-owned company. At the start of the project the coke plant and the rest of the Dunaferr operations were undergoing major changes because of the country's transition to a market-based economy. Due to this, Dunaferr became divided into more autonomous units. Isolated profit centres, as opposed to centralized planning, facilitated the process of problem identification in the various units. Also most of the operating and management practices were undergoing major changes. The financial problems of the company were exacerbated by the unrest in former Yugoslavia, one of its largest markets. However, managers was very interested in and motivated to learn Western management principles because they were interested in exporting to the West.

The coke plant produces approximately 1 million tonnes of coke from some 1.4 million tonnes of coal. The two coking units, based on Soviet technology, were built in 1950 and 1985 respectively.

The major environmental problems associated with coke production are primarily the leakage of crude chamber gas from the ovens, pollution from the outdoor storage of coal and from the grinding and crushing of coal. The industry is characterized by neglect and has the typical sign of a state planned economy, i.e. serious ground contamination.

The QEMS

The development of the QEMS started with the establishment of a checklist incorporating the requirements of the two standards. The checklist was a tool for the auditor assessing an organization against the two standards, but it also gave direction to management on what measures were necessary to provide assurance of acceptable performance. The QEMS has 17 "elements" divided into three sections.

The management part assesses what the organization has done to manage its processes and reduce the possibility for losses. Typical issues covered are the policies and goals of the organization, how legislative requirements are observed and how the impacts of the operations are assessed and registered. Organizational structures, roles and responsibilities and training are also covered. Other important aspects of the management system are 75 records and other documentation and a system description or manual.

The operation part covers the management and control of the value-adding process, i.e. coke production. Central elements are the formalized routines for contract reviews, design and process operation as well as handling, storage and packaging.

Finally, the control and improvement part deals with how non-conformities and process operations experiences are used for improvements. Continual, documented improvements are the underlying message of the two standards, hence this section deals with testing and inspection, how to address non-conformities and implement corrective actions and how to audit the system, which are all tools for finding opportunities for improvement.

Project activities

The main activities of the project have been:

- Development of a checklist based on the requirements of the two standards.
- Performing a QEMS audit on the selected division.
- Performing an assessment of the quality and environmental situation.
- Writing an audit report and reports on the quality and environmental status.
- Development of a QEMS implementation manual.
- Top management training in the principles of quality and environmental management and implementation.
- Internal work at the plant on implementation of the QEMS.
- Review and input to the QEMS system.
- Dissemination seminar for the company's other divisions.

Project findings and conclusions

A major problem was the lack of an existing management system which could provide management with the necessary information on the status of the division's quality and environmental performance. The existing management structures were bureaucratic and inappropriate to meet the new challenges the company was facing. The quantitative information on pollution and corresponding effects was not known.

An environmental investigation was therefore needed and was similar to that undertaken in a cleaner production assessment.

The division of responsibilities among the departments and among individuals was very unclear. Responsibilities for quality and environmental matters were, to a great extent, not allocated. Management representatives were appointed with responsibilities for quality and environmental protection. By describing the management system, roles and responsibilities have been further defined. However, this is not an easy task in Central and Eastern Europe.

Generally, understanding and awareness of quality and environmental problems on the shopfloor were very poor. Training was given a low priority and was primarily given to managers. Everybody's role in achieving goals on quality and environmental matters was poorly understood.

These issues are now central to the management system. Market pressures, particularly for quality systems, are forcing management to document their activities and achievement. Management culture in Central and Eastern European (CEE) countries is still fairly authoritative and it will take time before new ideas are disseminated down to the shopfloor. It is also recognized that improvement of safety and health is needed to foster a sense of care for the external environment. Getting employees interested in improvement work necessitates training and communication from the top.

CONCLUSIONS

A number of conclusions can be drawn from the Norwegian and Hungarian experiences described in this chapter.

First, the successful introduction of CP programmes in industry requires financial support from the government or other sources. This support appears to be necessary in countries with well-established environmental enforcement powers and those which are in the process of developing such powers.

Second, CP programmes are cost-effective and have long-lasting impacts in industry if:

- CP capacity-building programmes are introduced and the CP methodology is well understood;
- top management commitment to CP assessments is achieved;
- workers at all levels are involved in implementing the CP assessments;
- suitable management systems for continual improvement in environmental performance are established.

Finally, quality and environmental management systems standards are increasingly being required by the market. Their implementation in CEE countries is increasing, driven mainly by the requirements of the export market. The integration of management systems which combine quality and environmental matters seems to meet the needs of industry in the CEE region.

Section VI
Practical Case Studies from Smaller Companies

32
Introduction

RUTH HILLARY

The vast majority of businesses in the world are small and medium-sized enterprises (SMEs), for example in the UK in 1993 94.4% of all enterprises employed less than 10 people (Hillary 1995). SMEs' collective environmental impact is significant and so their adoption of sound environmental management and production methods is essential for progress towards sustainability. It is therefore appropriate that the practical experiences of SMEs is considered and is the subject of this, the final section of *Environmental Management Systems and Cleaner Production*.

In four chapters, this section discusses the key factors which motivate and support smaller firms to address their environmental impacts (Chapter 33); the management commitment of one small Italian chemical company to voluntary environmental performance improvements and transparent communications with its stakeholders (Chapter 34); the effectiveness of an aluminium casting company's management system in the face of a potentially major incident (Chapter 35) and the implementation of the EU Eco-management and Audit Scheme (EMAS), the British standard BS7750 and the international standard ISO 14001 at a high-risk medium-sized manufacturing site (Chapter 36). Very few small enterprises have made the positive link between good environmental management practice and good business practice. Internal and external factors inhibit SMEs from taking on new responsibilities; however, the core theme of this section is how smaller enterprises can adopt positive environmental actions.

In Chapter 33, Michael Smith provides a profile of the UK SME sector, discussing the key factors which stimulate smaller firms to address environmental issues. He argues that we are all stakeholders in the SME sector and as such the key for all stakeholders is to assist SMEs in identifying the opportunities to be gained from improving environmental performance. Smith proposes three levels of assistance: motivation, support and solutions. He asserts that smaller enterprises respond best if they are assisted through a partnership approach, citing the Groundwork partnership approach as a model. Three main features—cost savings, legislative compliance and

---

Environmental Management Systems and Cleaner Production, edited by R. Hillary.
© 1997 John Wiley & Sons Ltd.

market profile—are identified by Smith as motivating SMEs to address environmental issues, and he presents a selection of costed company examples to illustrate the point. Smith concludes that assisting SMEs to achieve quantifiable and sustainable results from environmental improvements is fundamental to demonstrating the positive link between environmental concerns and business.

Vittorio Biondi and Marco Frey, in Chapter 34, focus on one small Italian chemical company, Lati Industria Termoplastica, which has made the positive link between business and environmental performance highlighted by Smith. The authors discuss Lati's implementation of and certification to the British environmental management system (EMS) standard BS 7750 and its participation in the Eco-management and Audit Scheme (EMAS). Biondi and Frey suggest that Lati's participation in EMAS needs to be examined in the context of organizational and management transformation. The authors show how the company's top management view voluntary environmental performance improvements and transparent communication with the public as consistent with the company's total quality strategy. The authors assert that Lati's residential location is the reason for the company having adopted a proactive and transparent stance towards its environment performance. Biondi and Frey conclude that Lati is not representative of the average Italian SME, suggesting that the company's technological and organizational innovation describes its atypical behaviour.

Chris Burleigh, in Chapter 35, provides a detailed case study of a medium-sized aluminium casting company's improvements in production through the introduction of an EMS. Burleigh profiles the company's progress towards ISO 14001 certification, highlighting a potentially major environmental incident which tested the effectiveness of the system. He discusses the substantial cost savings and production improvements identified by the company's capability studies, asserting that the company's EMS procedures, in particular the communication procedure, played an important role in realizing the benefits. In conclusion, Burleigh questions the ability of the company to maintain the impressive cost savings already achieved, but suggests that the benefit of environmental management systems is their preventive capacity which helps companies avoid added costs.

In the final chapter and case study, Ken Jordan considers the practical experience of implementing EMAS at a "top-tier" CIMAH manufacturing site. He discusses in detail the stages of implementing the environmental management system standard BS 7750 at the medium-sized site, pointing out the importance of environmental awareness training of employees in the early stages of implementation. Jordan presents a method for determining significant environmental effects, arguing that the register of environmental effects is the heart of any environmental management system. He asserts that systems integration of both quality (ISO 9002) and environment (BS7750, ISO 14001) into the site's overall management systems is logical because their structures are similar and integration streamlines the internal auditing functions. Jordan concludes that the management of a site's environmental issues must be an integral part of any site's operations and that such management also makes very sound financial sense.

What the industrial experience in this section shows is that SMEs can systematically manage the environmental aspects of their activities. The successful implementation of environmental management systems in SMEs is a reccurring theme discussed in this section, but critical analysis by the authors shows that these case studies are atypical. The typical SME is reactive to the threats and opportunities presented by environmental issues, whereas larger firms have the capacity to respond even if they remain inactive. A common issue highlighted by this section is that smaller enterprises need powerful internal or external stakeholder pressures to trigger them into action. Since SMEs are the vast majority of businesses in the world, it is finding these triggers which is important to all stakeholders if SMEs are to participate in our collective progress towards sustainable development.

REFERENCE

Hillary, R. (1995) *Small Firms and the Environment*, Birmingham, Groundwork.

33
Stimulating environmental action in small to medium-sized enterprises

MICHAEL SMITH

INTRODUCTION

This chapter discusses the practical experience from environmental projects and initiatives undertaken over the past four years with small and medium-sized enterprises (SMEs) in the UK. The projects and initiatives have not only recognized the importance of raising awareness of environmental issues in the SME sector, but also successfully addressed the recognized problems in convincing organizations of their relevance. Even more importantly, the initiatives have demonstrated that, with the right approach, it is possible to provide SMEs with a framework that successfully delivers cost-effective environmental support services at a local level.

THE IMPORTANCE OF THE SME SECTOR

Communities are forced to cope with a variety of social, economic and environmental impacts. Those communities with declining traditional manufacturing industries often have an unwelcome legacy from these activities in the form of environmental degradation. In addition to the direct negative effect of the loss of jobs, the poor

---

Environmental Management Systems and Cleaner Production, edited by R. Hillary.
© 1997 John Wiley & Sons Ltd.

environment can be a threat to that community's ability to retain its workforce and to attract new business. The decline of larger operations naturally puts a new focus on existing small and medium-sized enterprises' potential to create new jobs. Any replacement employment generated is almost certain also to affect the ratio in favour of SMEs.

A few facts show how the UK's SME sector is a crucial part of the current economic picture. Whether traditional industries in the service sector or sunrise technology based, they form an increasingly important part of the environmental equation on a local and national basis.

UK's SMEs:

- are the largest sector comprising 2 686 500 companies at the end of 1992;
- are usually closely integrated into the fabric of the local community;
- have staff who usually come from within a small radius of the company;
- often use traditional processes or services;
- are often forward looking and dynamic.

We can probably all think of an SME round the corner from home. If it is a noisy, smelly paint-spray garage we may not look on our SME with too much affection—after all, would someone operating in a bad environment be expected to provide a good service or to be environmentally responsible? However, if that same garage looked tidier and thus likely to provide a responsible, reliable, trustworthy and reasonably priced service, it would become a very handy SME.

In some areas, such as the Black Country in the UK, a traditional SME has been the foundry producing brass badges and fireplace ornaments. The Environmental Protection Act 1990 is regarded as a major obstacle, in some cases legitimately, it is not seen as an opportunity. Although they normally only employ between two and five people, should such foundries close the knock-on effect can be quite profound to a local community, because they rent facilities, buy scrap and generate local income. They also, of course, pay taxes.

Those SMEs in the service or new technology businesses may not believe that they operate in areas that have any environmental impact. Opportunities may again be missed.

A good example of this is found in Taylors, a firm of commercial solicitors in Blackburn, who invested in an environmental review by Groundwork consultants, confident that their brand new premises were environmentally sound.

The immediate result of Groundwork's advice was a 33% reduction in electricity bills, saved without further financial investment.

We are all stakeholders of the SME sector. The key for all of us as stakeholders is to help the sector identify the opportunities presented by improving environmental performance.

PARTNERSHIPS TO ASSIST SMEs

In order to protect the local environment and strengthen the tax base, local authorities have a dual interest in helping firms to improve their internal management practices and enhance the investment appeal of the local community. However, local governments themselves may have neither the resources nor the technical expertise to accomplish this task. In these cases, broad-based partnerships with non-municipal organizations that provide environmental support services and access to additional funding can result in benefits to the local economy, the environment and the community at large.

THE GROUNDWORK APPROACH

Groundwork seeks to involve all parties—local and national government, community members and, in particular, business enterprises—in the improvement and maintenance of the economic and environmental health of their community. It seeks to remediate and regenerate the local environment to create favourable conditions more attractive to incoming business and better for the existing community. Groundwork also seeks to support small and medium-sized enterprises in complying with environmental legislation, increasing efficiency and profitability through the provision of diverse environmental management services.

An excellent example of this type of partnership is provided by the Blackburn Partnership in Lancashire. In 1988 the Blackburn Partnership Charter was signed with the interest and support of HRH Prince Charles.

The aims of the Blackburn Partnership are:

- To secure economic regeneration, increase economic activity and reduce unemployment for the benefit of all local people of the Borough of Blackburn.
- To seek a cooperative and coordinated working relationship between Blackburn Borough Council, the private sector, trade unions and the voluntary sector within the community.
- To build on the strengths of each of the partners within the partnership.
- To secure the involvement and support of the business community for the economic regeneration of the borough.
- To contribute to Blackburn Borough Council's annual economic development plan.
- To improve the quality of the local environment.

Groundwork in Blackburn was established as an integral part of the partnership, charged with environmental regeneration, winning hearts and minds and improving the environmental performance of business, particularly the 1400 SMEs in the borough.

The key partners are:

- Blackburn Borough Council
- Blackburn and District Council for Voluntary Services
- Blackburn Trades Council
- East Lancashire Chamber of Commerce and Industry
- East Lancashire Training and Enterprise Council
- Groundwork Blackburn
- Lancashire Council of Mosques
- Lancashire Evening Telegraph
- Local businesses
- Political party leaders

BARRIERS TO IMPROVING SME ENVIRONMENTAL PERFORMANCE

SME owners and managers cite a whole raft of reasons why it is important for them not to address environmental issues related to their businesses. A list of typical reasons why they should not address environmental issues follows:

- I'm a small or medium-sized enterprise: "what environmental effects?"
- You should see what . . . (multinational) gets away with!
- If I'm caught I'll plead ignorance and then do something, perhaps.
- I won't make a difference, when everybody else does something then I will.
- I'm behind on production, my client does not like a delivery, my assistant is ill, I had to mend our main machine last night and my best foreman lost his left finger yesterday. I'm too busy to deal with quality, health and err, umm, environmental issues.
- Consultants are for large companies with lots of money.

GROUNDWORK'S EXPERIENCE IN ASSISTING SME's IN IMPROVING ENVIRONMENTAL PERFORMANCE

Groundwork has taken a very proactive approach to contacting local businesses with support from its partners. This literally means personal calls to every business: very hard and expensive, but ultimately rewarding.

> Park Products, a manufacturing company with 65 employees, was not initially receptive to Groundwork's approaches.
>
> After many months of putting the issue on the back burner, we finally agreed to a meeting.
> We were surprised to find that Groundwork's advice brought us several immediate benefits including accessing grants and achieving legislative compliance—with more benefits such as reduced energy bills in the pipeline. (Paul Earnshaw, Manufacturing Director)

Groundwork provides three key levels of assistance:

- motivation
- support
- solutions

To **motivate** an SME, the main features stand out:

- cost savings
- legislative compliance
- market profile

In general, 10% savings on waste disposal and energy costs are typical. A small selection of possibilities are detailed to motivate involvement.

Surveys in Blackburn and Wigan have found that the majority of SMEs have only achieved 50% legislative compliance, specifically, for example, 30% complied with the Duty of Care Regulations, 35% complied with the Control of Substances Hazardous to Health Regulations and 40% complied with the Electricity at Work Regulations. There is now a selection of examples of fines imposed on businesses, which can be used to demonstrate the necessity of improving environmental performance.

> We were known as that mucky factory on top of the hill—where you probably went for a job as a last resort. We are now a far more attractive place to work and more attractive to potential employees.
> It's also far easier to generate quality consciousness in customer service and production processes when you work in a quality environment.
> Staff and customers now view us in a more positive light. (Paul Rink, Wolstenholme International)

There are also marvellous examples of the environmental profile of companies assisting in attaining greater market profile, but often the local community perception is also very important.

To **support** an SME:

- grants
- personality in whom to place trust
- information
- signposting
- practice transfer
- workgroups
- "buddy system"

Without doubt, grants make the first steps less painful, but ideally offer support to SMEs to a specific defined standard point. Blackburn has a grant for an environmental review and for implementing the review findings, whether BS7750 or a new heating system. The person offering the grant has a great deal to do with the success of that grant. They need to be independent, serious and sincere: Someone you can trust.

Information is provided in simple, concise English, quickly and in steady, bite-sized chunks. A 24-hour helpline with 48-hour response, monthly newsletter and the best environmental library in Lancashire are offered by the Blackburn Business Environment Association (BEA). If the BEA does not know the answer or cannot provide the solution the staff know someone who does—quick, independent signposting.

Every year Groundwork works with hundreds of different companies and has access to the experience of its major national sponsors, such as BP, Shell, Esso, RTZ, ICI, UK Waste, the Post Office and British Gas. Best practice is transferred with permission. A particular success has been in bringing small workgroups together to tackle specific issues such as the Eco-management and Audit Scheme (EMAS) regulation, waste minimization and energy monitoring. The buddy system links a sponsor or a major company involved with Groundwork to a recipient SME. Both parties benefit, but particularly the SME.

To provide **solutions** to improving the environmental performance of the SME, a wider range of services needs to be provided, ideally by staff/individuals/consultancies that advocate the Groundwork and partnership approach. The maximum benefit must remain with the SME to maximize the potential for economic and environmental regeneration. The solutions are:

- Checks
- best practice reviews
- audits
- training
- BS 7750
- EMAS regulation
- ISO 14001
- health and safety

- quality
- total quality management
- integrated management systems
- landscape management
- workplace green groups
- community projects

Some of the highlights of this solutions package are worthy of note. British Petroleum has undertaken to provide training to Groundwork staff to enable them to conduct reviews of SMEs. UK Waste Aware is a campaign providing information on Duty of Care and a free site review of the regulation. This supports IMPEL, the Department of Trade and Industry's *Information Manual and Programme of Environmental Legislation*. Workplace green groups are allocated a company budget, which normally attracts external grants, are trained by Groundwork and then implement site landscape improvements and subsequently participate in community landscape improvements.

> ICI Acrylics in Darwen funded staff to form a "green group". They began by carrying out improvements on their site and in areas surrounding their plant.
>
> The group has subsequently developed projects with local schools and community groups.
>
> The initiative has been good for staff morale and has helped cement local community relationships. (Kevin Leith, European Human Resources Manager)

Behind every solution, however, is the winning of the hearts and minds of those involved. Groundwork does this by illustrating individual potential to contribute to reducing serious environmental issues. For instance, if each car driver in the UK improved fuel efficiency by 1 mile per gallon, 500 million gallons of petrol would be saved. If each photocopier in the UK used 5 sheets less per day, 2.5 million reams of paper would be saved, 200 000 trees left standing and 3 million cubic feet of landfill left empty.

CASE STUDIES: THE FACTS AND FIGURES OF SUCCESS

No journey towards improved environmental performance is better stimulated than by practical examples, particularly when the owner/managing director is willing to talk to others. Here are just two:

A small company employing 20 staff operating an office and a warehouse was confident of full legislative compliance; it was positive that overheads were as low as possible and felt that it was doing Groundwork a favour by allowing the review to be conducted. The findings and benefits? Twelve areas of legislative non-compliance, waste management (£650/year), landscape improvements and reduced maintenance (£805/year) and improved transport control (£860/year). Nearly £3000 of savings each year, at an initial cost of £300 for the review. Subsequently President of the Board of Trade Michael Heseltine visited the company premises and the managing director enjoyed a trip to the House of Commons, courtesy of Groundwork and the Blackburn Partnership. Progress is being made towards BS 7750 and EMAS.

A papermill in Lancashire employs 160 staff, manufacturing recycled paper. The entire workforce has been provided with detailed environmental training, 28 staff form the environmental management action group (EMAG), legislative compliance has been reached and the following savings accrued:

Effluent recycling and water use reduction	£150 000
Energy management	£50 000
Pallet management	£60 000
Stock rotation	£30 000
Pulp reduction	£10 000
Oil recycling	£3000
QC laboratory tap	£1000
Total	£303 000

The mill has a site nature conservation plan, improved staff amenities and school links and is now a Groundwork sponsor and a board member.

CONCLUSION

Essential to the success of partnerships such as Groundwork and the Blackburn Partnership are the direct participation of local government, the shared commitment of senior management from the local business community, the participation of an independent service provider, access to diverse funding sources, and a professional, bottom-line orientation towards business.

Helping business leaders achieve quantifiable and sustainable results based on their own interests is fundamental to demonstrating the positive link between environmental concerns and business. This will result in both management and staff adopting a more responsible attitude towards environmental management practices in the company and at home. This change in attitude will produce widespread, long-term benefits for the entire community, including increased pride and morale.

34
EMAS adoption by an SME in the chemical sector

VITTORIO BIONDI AND MARCO FREY[1]

INTRODUCTION

This case study focuses on a small Italian company, Lati Industria Termoplastica Inc, which decided to control and improve its environmental impact by implementing an environmental management system. Lati's top management is convinced that a holistic approach to prevention is necessary to obtain systematic improvement of environmental performance. It decided to give a positive start to this approach by obtaining environmental certification (based on BS 7750) and participating in EMAS.

The EMAS (Eco-management and Audit Scheme) regulation (no.1836/1993) introduces a new instrument based on a voluntary approach in the European Union (EU); this tool is consistent with the Fifth EU Action Plan on environmental policy which is dedicated to sustainable development. The EU decided to promote collaboration between companies, institutions and the public, with the aim of improving environmental quality. Enterprises are encouraged to take on environmental issues according to a voluntary approach which will gradually substituting for the traditional "command-and-control" approach.

A company that wants to participate in EMAS must organize an environmental management system oriented to continuous improvement. This system includes a sequence of management steps: initial review, policy, programmes, activity implementation, audit and statement. Company commitment must be documented and validated by an external verifier. Initiatives, objectives and methods of action are left to the company.

Environmental Management Systems and Cleaner Production, edited by R. Hillary.
© 1997 John Wiley & Sons Ltd.

EMAS aims not only at favouring and diffusing positive and proactive actions in the industrial system, but also at stimulating external communication and a transparent relationship between the company and its stakeholders. EMAS requires systematic interaction between the production site, the public authorities and the population living around the industrial location.

In Italy, it is difficult to find companies which are open to interaction with external stakeholders. The transparency of information is greatly limited due to difficult relations between companies and both authorities and the general population. An economic characteristic of Italy is the large number of small and medium-sized enterprises (SMEs) which constitute the core of the economy. In the Italian chemicals sector, for example, 40.8% of companies have less than 20 employees and 40.4% have less than 100. In the last few years, exports have strongly increased for all in the chemical sector and for SMEs. For companies with less than 500 employees, the percentage of exports in the total amount of production increased per employees from 27.5 (1991) to 31.8% (1993).

Chemical companies are beginning to feel competitive pressure from customers in the area of environmental quality. A high level of environmental performance will become more and more important for business competitiveness in the EU. SMEs are confronted with many organizational, financial, cultural and technical difficulties when implementing environmental management systems. It is very important for the public authorities to support their initiatives by financial and other incentives (such as continual technical, legal and training assistance).

LATI: COMPANY DESCRIPTION

Lati Industria Termoplastica Inc, founded in 1945, is a chemical company located in Vedano Olona, a town of 7000 inhabitants in the Varese province, about 40 kilometres from Milan. In the early 1960s it was the first European company to produce glass-fibre reinforced thermoplastics. Today Lati has 250 employees, with a turnover in 1995 of 163 billion lira for a production of 24 000 tons.

In Table 34.1 data are presented related to turnover, general and environmental investments for the company in the last five years. Environmental investments are incorporated within the general investments, but there is a specific quote of turnover (< 1%) strictly allocated to environmental investments.

Lati recently built a new facility in Torba, a village located a few kilometres from Vedano Olona. At this site, the top management adopted the best technical solutions to control environmental and safety issues. Lati supplies most of the thermoplastic compounds utilizer industries, i.e. cars, electrical appliances, electronics, informatics and mechanics. The company exports more than 50% of its turnover to Austria, Germany, France, the UK, Spain, Switzerland, the Benelux and Scandinavian countries and Brazil.

At Vedano Olona, Lati carries out research, development, production and distribution of thermoplastic compounds. The site is split into seven areas: warehouse,

Table 34.1 Turnover, investments and investment in environmental protection (million Italian lira)

Year	Turnover	Investments	Environmental investments
1991	73 000	7000	200
1992	74 500	6770	250
1993	78 600	6100	170
1994	112 000	9800	300
1995	163 000	12 000	752

additives and colouring matering, base mix preparation, thermoplastic materials extrusion, driers, homogenization and packaging, and logistics and final products. The processes are centrally regulated by a mainframe computer which plans and monitors overtime production phases.

Each product department presents different features in organization, technology, management and environmental effects. In addition to these areas, the site has a laboratory where very important company activities are carried out, including the definition of compound formulae and production methods to meet customer specifications. The characteristics of each product lot are illustrated by an analysis bulletin. As discussed later, the company, through its environmental review, has analysed each phase of the production process, identifying the environmental impacts and planning specific actions for impact minimization.

Particular attention is given to customer technical assistance in new project development. Lati tries to anticipate customer needs through continual product innovation. Research and product optimization are the main objectives of the corporate strategy.

QUALITY AND ENVIRONMENTAL MANAGEMENT

In the mid 1980s, Lati began to reorganize its activities with the aim of realizing a business strategy oriented to total quality management. Starting from a clear commitment from top management, this process led to a detailed analysis of Lati's organizational structure and procedure definition. About 60 organizational procedures and 400 operative procedures were written or modified and training programmes involving all employees and company activities were set up. The change was not limited to organizational innovation. Top management redefined the planning process and completed the automation of different departments using an optical-fibre network. An important step in the process of management improvement was achieving quality certification to ISO 9001 in March 1993.

In operational terms, the main objectives achieved in this first phase of reorganization were:

- definition of supplier product characteristics;
- development of a maintenance plan;
- instrument calibration and control plan;
- quality function deployment (which in the last two years has produced a significant reduction in customer complaints);
- full document retrieval and traceability;
- periodic supplier evaluation;
- periodic internal audit;
- definition of product characteristics on customer specifications.

Lati's participation in EMAS must be examined in the context of organizational and management transformation. The company's top management considers the process of voluntary environmental performance improvement and transparent communication with the public to be consistent with Lati's total quality corporate strategy. In this light, legal compliance is a prerequisite for environmental management, not the main objective. Great importance was given to the capacity continuously to maintain compliance within the perspective of continual improvement. Consequently, Lati allocated specific resources for continual monitoring, management control and the review of company objectives.

At the same time, Lati's management believes that it is important to link company objectives with stakeholder requests. Lati understands that the different stakeholders involved in environmental issues are more than those involved in quality management. While in quality management the most important stakeholders are suppliers and customers, in environmental management other players must be added: public authorities, regulators, local community living around the site, non-governmental organizations (NGOs), banks and insurance companies. Each player has its own special rules to the game, language, requests and information needs; therefore it is difficult for the company to identify how and what to communicate. It could be useful to select the most strategic stakeholder to whom the environmental management system and information are addressed. This crucial issue is discussed in greater depth below.

Lati decided to initiate the process leading towards EMAS implementation in the early months of 1994, when the entire regulation was not yet operational in institutional and legislative terms. Top management was acquainted with the difficulty of rationalizing the organization and its activities to improve environmental performances. The authors worked with the company during a pilot project on the diffusion of EMAS in chemical SMEs cofinanced by the European Commission (DG XI). The pilot project finished in November 1995 and Lati is at an advanced stage of EMAS implementation, possible one of the most advanced Italian SMEs in the EMAS process.

INITIAL ENVIRONMENTAL REVIEW

Lati's initial environmental review took into consideration all the environmental effects associated with the activities of the Vedano Olona site in order to identify the most significant ones. According to the EMAS Regulation (Annex I.B.3), the review was performed taking into account normal and abnormal operating conditions; incidents, accidents and potential emergency situations; past, current and planned activities.

The most important environmental effects identified by performing the initial environmental review were:

- dust and noise emissions within the warehouse and in the metering, base mix preparation and extrusion areas;
- washing water and degassing water production;
- production of urban, special and toxic/noxious wastes (as classified by Italian legislation);
- bad smell emissions.

After examining the environmental effects linked to the activities of the site, a register of the environmental effects was drawn up (Annex I.B.3). The results of the environmental review represent the basis from which the systematic compilation of the register and the subsequent assessment of future changes to effects was established.

ENVIRONMENTAL POLICY AND PROGRAMMES

The main principles of Lati's environmental policy are the following:

- to comply with all legal requirements applicable to the site;
- to protect the environment;
- to respect the physical and mental integrity of human beings.

On the basis of these principles, the company decided to prepare periodical improvement plans with the goal of evaluating the consequences for the EMS of each process/product innovation. After these evaluations, the EMS manager prepares specific action plans to ensure that there are monitoring and control of all the environmental effects connected with the different production aspects. The manager submits these action plans to top management to enable all the environmental aspects connected with each project to be accurately evaluated.

The most interesting characteristic of Lati's environmental policy is its close link with the entire planning process of the company, i.e. it is very important for top management to take decisions about process/product innovations which also take into account the opinion of the EMS manager. Without this link, the environmental policy is destined to become just a formal declaration of principles.

After the initial environmental review, Lati undertook the following actions:

- dust exhaust bag filter and fume hood installation in all areas with dust production, in addition to dedicated dust exhaust equipment for the metering department;
- soundproofing protection of the most noisy equipment (delivering booster toward silos, cutting machines, cyclones installed on extrusion lines, conveyment piping for storaged products etc.);
- installation of two treatment plants for industrial water coming from the production and cleaning activities (analysis of water before and after treatment is performed on a monthly basis);
- installation of a degassing water treatment plant: waste water coming from this plant is stored as special waste, but through a dedicated physical chemistry plant it is treated and returned to the closed circuit;
- all wastes are stored in dedicated areas with waterproof bases, wastes are packaged in order to avoid any dispersion both to the environment and to the ground. No significant accidents have happened in the past, but still prevention measures have been adopted to avoid incidental losses to the ground of noxious classified substances;
- the smoke deodorization plant completed in 1990 is presently the biggest in Lati. All emissions from the production lines are conveyed towards a single centralized pipe for exhaust air collection, and sent to the deodorization plant, after sodium hypochlorite oxidation, when it is possible to send the emission to the air.

All these actions enabled Lati largely to comply with current environmental legislation and to pursue the continual improvement of its environmental performance. Nevertheless, in order to realize this improvement, Lati decided to strengthen its efforts with the aim of reducing odours connected with its activity. In the past, odour had created several problems with the company's relationship with the local community. The site is located in a very densely populated area. In the 1950s, at the beginning of the site's activity, the area was an agricultural zone, but the small town grew quickly during the 1960s and 1970s. This fact makes the local population pay a great deal of attention to Lati's activities and the company has to monitory continually the odours produced by its activities. This is why, in 1990, Lati decided to invest a large amount of money in the installation of a new deodorizing plant.

EMS IMPLEMENTATION

After carrying out the initial environmental review and defining its policy and programmes for the most significant environmental effects, Lati had to define, organize and implement its EMS, the main goal of which is to realize the environmental programme. Following the definition given by EMAS, Lati designed its EMS tailored to the needs and characteristics of the site.

From the organizational point of view, there is an environmental committee composed of a representative of top managers, the EMS manager, the production manager, the company doctor, the personnel manager and three representatives of the employees. Lati clearly defines the responsibilities and resources of the EMS manager in strict relationship with top management. His/her main tasks are:

- to guarantee compliance with all the environmental legislation relevant to the site;
- to draw up the annual action plan to realize the environmental policy;
- to write and diffuse environmental procedures;
- to organize and conduct periodical internal audits;
- to prepare (in collaboration with the internal medical service) the annual safety plan;
- to write the environmental statement;
- to coordinate the activities of the environmental committee.

Moreover, there are other tasks and responsibilities connected with environmental management. In particular:

- top management defines the policy, decides the programmes, receives the audit reports and approves the statement;
- the production manager is responsible for the environmental performance of the site.

Direct involvement of top management, and not only of the EMS manager, in the environmental management of the site is a necessary condition to enable the EMS to reach the company's objectives. Because of this, Lati is able to:

- involve all the employees actively in the environmental management system, keeping them informed about technical, organizational and legal issues;
- guarantee the necessary balance between the responsibilities and resources given to each department;
- continuously verify the adequacy and completeness of resources in relation to corporate objectives;
- encourage the spontaneous creation of working groups and committees to improve the diffusion of information and skills.

A relevant role in this approach is played by the computerization of information flows. Lati decided to become, by the end of 1999, a computer-integrated enterprise. This goal will be achieved by the following: the introduction of a fibreoptic network (1992–95), the computerization of the entire organization (more than 100 PCs are operating in the company) (1991–94), the creation of a geographic network involving all departments and the commercialization network (1993–99) and the complete automation of the manufacturing activities (1991–96).

In this organizational approach human resource management is crucial. The company prepared an internal training plan with the aim of satisfying the general training needs, individual needs and the training needs of new employees. All training activities have a specific part to play regarding the environmental management system, both general and specifically connected with individual tasks. Personnel training is one of the areas where the intergration between health, safety and environmental management is most fruitful for Lati.

Procedures and documentation management are comprehensive. Lati prepared the following documentation: environmental review, improvement plans, register of the environmental effects, EMS manual, organizational procedures, operating procedures, audit reports, instrument calibration reports, design reports, non-compliance reports, product safety schedules, maintenance reports, training reports, environmental effects of subcontractors and customers, and the environmental statement. This is a long and complex list. Each document has its own procedures for writing, collecting, updating and diffusion. The experience gained by Lati in the quality management was very useful in organizing its documentation for the EMS. It is necessary to document the EMS to help both the company's management and the verifier. Nevertheless, it is very important for the company to avoid excessive production of documentation that could become a serious problem and compromise the efficiency of the entire management system.

The last EMS element to consider is the auditing activity. According to the definition given by the EMAS regulation, Lati considered auditing the main tool for redefining and monitoring the efficiency of its EMS, for reformulating its environmental programmes and objectives and, if necessary, for modifying the EMS. Lati prepared specific working documents for conducting audits, established an audit plan and trained a group of internal auditors. It is very important to emphasize the fact that the audit teams directly refer their results to the company's top management, who can take all necessary actions from the audit.

Lati's decision to use the internal audit in this way cannot so easily be extended to other SMEs. The human and financial resources required for the implementation of an audit activity is not always within SMEs' reach, so they often use external consultants. Lati took the decision to use internal audits because of its strong approach to improving its internal human resources.

EXTERNAL COMMUNICATION

As already mentioned, the site is located in the middle of a very crowded zone. This is why the company decided to adopt a proactive and transparent approach to its environmental communication. Lati drafted and diffused the environmental statement paying particular attention to the two requirements defined by the EMAS regulation for the statement to be concise and clear. The statement was drafted on the basis of the environmental effects identified as significant by the initial environmental review.

The statement included the different activities performed in each department, with a simple description of the processes which take place in the plants and their associated environmental effects. A brief description was given of the environmental management system and of the environmental programme, including summarizing, clearly and exhaustively, the practices, procedures, processes, responsibilities and resources necessary for implementation of the programme.

The statement is addressed to the "public". Lati identified its most important public or stakeholders as employees, the local community, the local health council and regulatory bodies, customers and suppliers. The company took due account of the widely differing technical skills, scientific knowledge and environmental culture of each of these stakeholders, trying to give all the relevant information in the statement but at the same time avoid adopting technical and specialist terms.

CONCLUSIONS

The company analysed in this chapter is not representative of the average cultural and managerial level of Italian SMEs. Historical, economic and technical reasons have led Lati to adopt a proactive management approach. Flexibility, technological and organizational innovation, adoption of cleaner technologies, the pursuit of continual performance improvement and globalization of its marketing approach are a few of the key approaches which explain Lati's behaviour. They also indicate why the company decided to adopt a proactive approach towards environmental management.

The familiarity of the management with clearly defined responsibilities, involvement of all departments and employees in this process, data and information circulation and continual monitoring of performance facilitated the EMS implementation. The decision to pursue integration between the quality system and the environmental management system is another interesting point emerging from this case study. When a company's management understands all the similarities and differences between quality systems and environmental management systems, they have the opportunity of exploiting many organizational and managerial synergies.

At the end of the process leading to EMAS registration, Lati identified the main benefits connected with the implementaton of an EMS. They were:

- higher capability in the planning activity;
- increases in environmental management efficiency;
- higher level of competitiveness (e.g. in relations with customer and suppliers);
- reduction in management costs (e.g. operational costs, energy consumption, risk assessment);
- improvement in the process of self-evaluating skills;
- improvement in the relationship with local stakeholders;
- development of the corporate environmental culture.

NOTE

1. The authors give special thanks to Maurizio Gilioli and Fabio Iraldo for their comments and suggestions.

35
Achieving improvements in production through environmental management systems

CHRIS BURLEIGH

INTRODUCTION

Participation in a pilot study during 1992–4 on the implications for small companies introducing BS 7750 and the Eco-management and Audit Scheme (EMAS), which resulted in promising initial results, led the managing director of an aluminium casting company to make the corporate decision, in the autumn of 1994, to pursue certification to BS 7750 and ISO 14001, once the international standard was adopted. Responsibility for its implementation was given to the group health and safety coordinator. This decision was based not only on the prospect of customers' future insistence on suppliers having legitimate environmental policies, but also on the expectation of improved productivity and efficiency. The case study that follows is an example of how this medium-sized organization has benefited from the introduction of an environmental management system (EMS).

Environmental Management Systems and Cleaner Production, edited by R. Hillary.
© 1997 John Wiley & Sons Ltd.

COMPANY PROFILE

The company employs some 400 personnel in the manufacture of high-volume quality aluminium castings, mainly for the automotive industry, from a gravity diecasting process based on three sites, two located in the West Midlands which produce small and medium-sized castings and the third in Wales which specializes in larger sizes.

The company had already achieved several coveted standards for quality management systems, including BS EN ISO 9002 and Ford Q1. Consequently, it was felt that to avoid duplication and confusion any additional management system elements such as BS 7750 and ISO 14001 should be designed to operate within the existing framework of documentation. On examination of the requirements of BS 7750 and the early drafts of ISO 14001 it was apparent that there was a lack of internal resource and environmental knowledge and so external skills were acquired through Target Environmental Systems, a specialist consultancy which had designed and implemented EMS for several blue-chip companies.

TRAINING

Through its work with other companies, Target had established that, while staff were often genuinely trying to be helpful in supplying environmental information, by not fully understanding what was being asked of them they sometimes glossed over or omitted vital facts important to establishing the EMS. Hence general awareness training was an essential starting point. Consequently, initial awareness and implementation training was started almost immediately. Training emphasized the lack of awareness in industry in general, a theme which was later to play a major part in one of the company's environmental initiatives.

A more detailed training programme was initiated after the company received approval from the local authority to operate a prescribed process and the managing director had approved, in February 1995, the initial draft of the integrated quality and environmental management system. Training included environmental auditing, which was to use a cross-section of employees from all three sites who were part of the company's existing quality auditing team.

TOWARDS CERTIFICATION

Fine tuning of the integrated management system and the addition of several new operating procedures and documents took place over the next few months, until the group health and safety coordinator decided that the time was right for the first step in the certification process and obtained verification for the system.

The company was on course for an initial assessment of its integrated quality and environmental management system by the British Standards Institution (BSI) in

November 1995. The final document, the register of environmental regulations, was introduced to the system and the company received an excellent report from the BSI assessor following the Phase 1 Assessment for BS 7750 for the graphic design and user-friendly format of the integrated quality and environmental management system.

POTENTIAL INCIDENT TESTS THE EMS

The following month was to see a potentially major incident that would test the effectiveness of the management system to the full.

On 7 December 1995, the production manager of the Welsh plant received a telephone call from the site manager of the local landfill site. He reported that the company's latest delivery of waste sand had been contaminated with a number of shiny blocks that glistened in the snow when they were observed during the tipping of the waste.

The waste sand arises from the gravity diecasting process. This process involves manufacturing an inner core made of sand, binding agents and catalysts which is positioned inside a die. Aluminium ingots are melted in the main furnace and poured into a holding crucible while it is transferred to the location of the die. The molten aluminium is then hand poured using a ladle into the die mould and allowed to cool. Once cooled the casting is removed from the die and the spent sand core is knocked out of the component, thus producing the desired internal features and dimensional characteristics with minimal machining requirements. The waste sand is then collected and sent to landfill via a waste contractor.

The contamination of the sand identified by the landfill site manager required investigation. The company's "emergency situations rapid response" procedure was immediately implemented, which included the generation of a "sequence of events report" to record all details and provide the necessary data for future analysis. The "emergency situations" procedure was one of the new procedures added to the existing management system. This is designed to address all potential major environmental or health and safety incidents with a view to identifying the immediate actions required in order to minimize the impact on the environment and remove the threat of further damage.

From the description given by the landfill site manager, the "shiny blocks" were obviously sodium, which is added to the molten aluminium while it is in the holding crucible to maintain and modify the grain structure in the casting. The potential for serious incident occurs when the sodium becomes damp. It may ignite, or more seriously, when sodium comes into direct contact with water it will almost certainly explode. Hence sodium at a landfill site in the middle of winter with snow on the ground was extremely dangerous.

Even though discovering the reason for sodium being disposed of in this extremely alarming manner was a major concern, especially considering the volatile nature of sodium and its "special waste" status was well publicised within the company, the immediate requirement was for the potentially explosive situation at the landfill site to be made safe. The landfill site manager was informed that the material was suspected to be sodium and given safe handling and storage instructions over the phone. Within 15 minutes of receiving the phone call the production manager, site superintendent and a foundry operative went directly to the landfill site with a container of paraffin which would provide a safe medium to store and transport the sodium.

On arrival at the landfill site the foundry personnel reported to the site manager's office, where they were informed that 56 25 g blocks had been discovered and placed in a box for safety as advised. A subsequent search of the landfill site was undertaken and a further three blocks of sodium were found. All 59 sodium blocks were made safe after being placed in paraffin and transported back to the foundry.

On their return, the production manager informed the local Environmental Health Department of the situation. This department announced its intention to send environmental health officers to make a visit to the foundry and conduct an interview into the incident. This was arranged for four days later, on 11 December.

Investigation and identification

Having successfully removed the potential for a serious accident, the next step in the procedure was to investigate and identify the causes and take preventive action.

Initial findings indicated that the skip had been part of a general clean-up operation to improve the visual appearance of the factory. All employees involved in this clean-up were interviewed to determine if they had accidentally or in ignorance placed the sodium in the sand skip. All employees reported that they were well aware of the safe storage, handling and disposal requirements of sodium due to the training they had received. Also none of the staff could recall placing any sodium in the skip by accident. Further investigation failed to identify the source of the contamination.

Following the meeting with the local Environmental Health Department a copy of the "sequence of events" report was taken by its representatives to confirm the facts. It was decided that, although the source of the contamination could not be identified, the site's emergency procedures had been effective in preventing a more serious outcome and so no further action was to be taken.

Although the local Environmental Health Department had been satisfied, the group health and safety coordinator was still concerned that there was the potential for a repeat incident, as the company had not closed the loop in preventing reccurrence. As part of the overall company policy of cascading information throughout the organization, details of the sodium contamination were made available at the other two sites in Birmingham. The findings of the investigation were also referred to the company's "improvement team" to determine further actions.

Another new procedure, "communications" was to play a small but invaluable part in the next stage of the process. This required all environmental information to be channelled through the group health and safety coordinator who, because of the register of environmental regulations, was also aware of legislation to phase out the use of sodium by the end of 1996. Following discussions at the other two plants, an employee made the suggestion that the sodium could possibly be replaced by strontium. Strontium had already been on trial at the sister plants for several weeks following a quality initiative and this suggestion triggered the "preliminary feasibility" procedure and was passed to the improvement team who were tasked with processing the initiative and undertaking further trials.

When this information reached the improvement team it was discovered that a second initiative was already underway, instigated by the improvement team themselves when they had decided to find an alternative to landfill disposal for the waste sand from the process.

Initial indications were that the strontium would be more expensive to operate than sodium due to higher raw material costs, but data from the other plants had shown that the effects of strontium were evident for nearly three times longer than sodium, and so a preliminary capability study was undertaken to determine actual results on the larger components.

Capability studies

The results of the capability study revealed some encouraging facts. Data from the study confirmed that the effect of the strontium was evident for three times as long as the sodium, so it was cost-effective to introduce strontium use as a substitute for sodium. In addition, using strontium resulted in a remarkable improvement in the quality of the products produced, i.e. the number of leaking castings was reduced by almost 30%.

Leaking castings have to be reworked using a process called impregnation, which involves subjecting each component to pre-wash, vacuum drying and cooling processes. Vacuum impregnation of a sealant and drainage operations follow and finally a cold water wash is given before a hot water cure. The company had been spending on average £20 000 per month on the impregnation process before the strontium was introduced, but the average costs have now been reduced to around £14 000 per month. Another benefit of strontium's introduction was improvements in the working environment for the workforce, who had for some time raised concerns about the use of sodium and its visible emissions.

One further benefit of strontium/sodium substitution was the increased working life of equipment. Due to the extremely aggressive nature of sodium the working life of the 70 holding crucibles within the company was originally between 12 and 14 weeks; on changing to strontium it was found to be extended to around 16–18 weeks. Each crucible costs between £265 and £380 depending on size, thus the annual saving to the company was in the region of £11 000.

While a satisfactory conclusion had been found and implemented for the initial problem of sodium contamination, the improvement team was finding it more difficult to reach a solution for the disposal of waste sand. Once again, the communication procedure was to play its part in giving the initiative a shot in the arm. The group health and safety coordinator received information from a major customer's environmental advisory department that landfill charges were set to rocket over the coming years. The company produces in the region of 5800 tonnes of waste sand every year, for which they were paying £6 per tonne for landfill. The anticipated charges for the first increase alone in 1996 were for £13 per tonne, raising annual charges from £35 000 to nearly £80 000.

During the course of the environmental effects and analysis procedure a concern was raised in the failure modes effects analysis (FMEA) over the toxicity of the sand due to the possibility of it leaching resin binders. Subsequent soil analysis proved this to be unfounded, but while the sand was found to be non-toxic the company decided to classify it as having "moderate environmental impact" due to the large amount generated. This classification, coupled with the forthcoming increase in landfill charges, helped to focus attention on finding an effective alternative to landfill.

Feedback from the supplier environmental assessment questionnaire, sent out to all suppliers of products and services as part of the vendor assessment procedure, was scrutinized with renewed emphasis on the use and disposal of sand. The review indicated that the sand supplier knew of an outlet for waste sand. When the supplier was contacted it was discovered that the supplier sold its own waste sand to various sectors of the construction industry. However, the improvement team's initial attempts to find an outlet for the waste sand were unsuccessful. They were repeatedly told by slab and brick manufacturers that they had a plentiful supply of cheap sand available and that they would not be interested even if the sand were to be given to them.

The setback may have proved too much for another organization, but the company was determined not to incur the additional landfill costs. Shortly afterwards and following discussions with its suppliers and contacts in other industries, a reclamation process was brought to the improvement team's attention. This process seemed to provide the ideal solution. It involved incinerating the sand to remove all traces of residue. The process was immediately put on trial at the company's Welsh plant. The results of the trials were judged to be excellent, with reclaimed sand as good as virgin material in both appearance and performance.

Unfortunately, there was a major obstacle in the way of the reclamation process. The component which had the biggest usage of sand cores originally had an expected life span of only two years, so with the price for the reclamation unit at £200 000 and additional running costs to take into account, best estimates put the payback period at around four years. Due to concern about the company's ability to sustain the level of usage the reclamation process was not viable, and so the search for an acceptable alternative continued.

Once again, the improvement team went back to the company's suppliers to dig deeper than ever and eventually, despite being told on several occasions that they were wasting their time, an end user was found within the construction industry.

Following initial discussions the potential "customer" was reluctant to pay for the sand and even refused to arrange or pay for collection, which looked initially to be unacceptable. However, following a detailed costing exercise, even accepting non-payment for collection was found to provide a net cost saving for the company of between £5000 and £10 000 per annum without taking the higher landfill charges into account.

With the initiative making progress, the following weeks saw various other ideas for using the sand in construction applications, of which three are now in the process of being investigated by the improvement team. While a final decision has yet to be made, the company is confident of saving even more money.

While this example illustrates how a potentially dangerous mistake can provide the impetus for a company to find better, cleaner solutions for production and save money in the process, the emphasis of the company's management system is on prevention rather than cure.

ENVIRONMENTAL INITIATIVES

Many other environmental initiatives have been instigated as a result of the BS 7750 programme, including the following.

An environmental audit revealed that many metal objects, including the metal straps for banding aluminium ingots together, were being placed in the general waste skip for which the company paid its waste contractor to dispose of to landfill. The company now collects all metal materials in a scrap metal skip for which it receives payment on the contents.

Following a major break-in at one of the Birmingham plants which resulted in £55 000 damages and loss of equipment, a visit from the local crime prevention officer revealed that the site had many hiding places for burglars and suggested an open-plan approach. This resulted in selling obsolete tooling which was stored on site, raising approximately £25 000. Further plans are in hand to sell off obsolete silos which will also improve the appearance of the site.

Another initiative was originally driven by one of the major customers who wanted to change from cardboard packaging to durable returnable packaging. The company approached its other customers who agreed to 50/50 funding in return for a piece price reduction.

The company's future initiatives and one currently being investigated include crushing the spent sand cores to reclaim aluminium fragments and flashings. While no accurate costings are yet available, with raw material costing £1200 per tonne and with approximately 500 000 castings a year, the expectation is for the payback period to make it a viable investment.

The company has now adopted Target's schematic diagram and flowchart format for its complete management system, which will greatly assist in training employees and also reduce time spent on maintaining the system. Plans for certification are scheduled for final assessment to ISO 14001 by BSI during the spring of 1997 and, with the guidance the company is receiving from Target Environmental Systems, it is confident of being successful at the first attempt.

In the months leading up to the sodium contamination incident the group health and safety coordinator had made repeated attempts to get one of the company's two waste contractors to complete a supplier environmental assessment questionnaire. He eventually gave up and granted the whole contract to the proactive waste contractor who was more than willing to share information and views. He now feels that this decision has been justified. He lacked confidence in the other company which was obviously being evasive. This led him to give his legal responsibility for the disposal of waste to a proactive organization that also employed the services of a legitimate landfill site. This proactive organization has not only helped the company avoid a major incident, but also saved them a large amount of money in the process. As the group health and safety co-ordinator himself has said, "For all I know, if I had kept the other company on we may still be waiting for someone to find the sodium, and that may have been a child somewhere!"

CONCLUSION

While the company has seen some impressive savings from its environmental management initiatives, there has been some discussion over whether these savings will be seen again. While nobody can foresee the future, one thing is firmly believed at every level within the company: the integrated quality and environmental management system will not necessarily identify huge ongoing cost savings, but it will help the company to avoid cost-added elements in the first place.

36

The practical implementation of BS7750, EMAS and ISO 14001 within a medium-sized manufacturing site

KEN JORDAN

INTRODUCTION TO AKZO NOBEL CHEMICALS' GILLINGHAM SITE

The site in Gillingham is part of the business unit polymer chemicals, within the chemicals group of Akzo Nobel. The site was officially opened as Novadel Ltd in January 1938 for the production of white lead, associated paint products and additives for the flour-milling industry. The 18-acre site on the banks of the river Medway in Kent is now one of four major organic peroxide producing locations within the EU operated by Akzo Nobel's chemicals group and employs some 140 personnel.

Five major manufacturing units, with several minor units, produce speciality chemicals including organic peroxides for the plastics and rubber industries and a monomer for the production of an organic glass for the optical industry. The site exports 95% of its manufactured tonnage outside the UK, with Europe taking two-thirds of the total production. The Middle East and Far East are fast-growing markets, with sales to these areas increasing from 10% of total sales to 20% in the last three years. The site also serves as a UK distribution centre for other Akzo Nobel products produced outside the UK.

Environmental Management Systems and Cleaner Production, edited by R. Hillary.
© 1997 John Wiley & Sons Ltd.

The site was registered to ISO 9002 in June 1990. A commitment to achieve the investors in people award was signed in June 1993, with registration in February 1996. The site was assessed for the British environmental management standard BS 7750 in March 1994, with accredited certification granted in March 1994 and registered to the European Union's Eco-management and Audit Scheme (EMAS) in August 1995. Gillingham was runner-up for the title of "Kent Company of the Year" in February 1994 and 1995, second in the "Kent Exporter of the Year" award in March 1994 and winner of the "Kent Business Award" for the environment in 1995. The site is also a signatory to the CIA Responsible Care Programme and a member of the CBI Environmental Business Forum.

The final step to EMAS registration was taken in June 1995 when the site's public environmental statement was verified by BVQI. In July application was made to the Department of the Environment and the site was registered to EMAS in August 1995. The site was also certified to the draft ISO (DIS) 14001 by BVQI in March 1996 achieving certification to the full standard in October 1996.

The site in Gillingham is a "top-tier" CIMAH site (control of industrial major accident hazards), due to the storage of over 200 tonnes of oganic peroxides. However, due to only a negligible amount of emissions of VOC and other prescribed substances, the site is not required to obtain authorization for its processes under integrated pollution control (IPC). This is a fairly unusual situation for a chemical manufacturing site and allowed the company to concentrate on the environmental management system BS 7750.

BS 7750 ENVIRONMENTAL MANAGEMENT SYSTEM

When planning the introduction of BS 7750 the company used the implementation circle from the front of the standard. However, this has since been modified to include training and the environmental statement required for EMAS (Figure 36.1). It started off with a commitment when the standard was launched in March 1992 and in March 1994 it received accredited certification to BS 7750. Akzo Nobel has a corporate policy to introduce environmental management systems (EMS) in all of their plants and when it compared Akzo Nobel systems with BS 7750, the standard was, at that time, superior; a recent update of the Akzo Nobel EMS has now brought it into line with EMAS and ISO 14001. However, an integral part of the Akzo Nobel EMS was a Dutch chemical industries environmental questionnaire. This questionnaire was used to carry out the initial environmental review. The questionnaire was modified to suit the particular circumstances of the Gillingham site's operations and was used for the initial review.

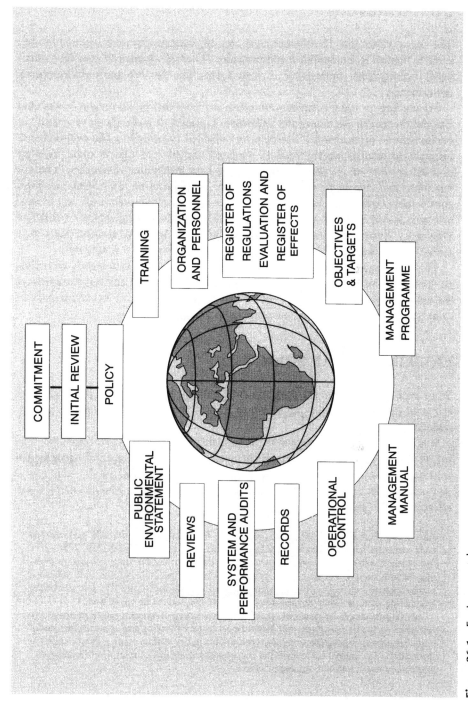

Figure 36.1 Environmental management system

INITIAL REVIEW

The site was split into 25 different areas and different people were assigned in these areas to complete the modified questionnaire. The questionnaires were then correlated together and summarized to give a base line for the site's environmental performance.

Several targets and objectives were identified from this initial review. It was clear that the site met all environmental legislation. However, it was difficult to quantify its environmental performance. The company decided to compile a site environmental information manual which could be updated annually so that it could maintain detailed records of its performance for air, land and water discharges. The site environmental information manual has made the writing of the EMAS statement considerably easier. The compilation of the manual took approximately six months and was carried out by two chemical engineering sandwich students from Bradford University. Systems were set up to monitor and correlate environmental data, on a monthly basis, such that the annual update of the manual was a simple task.

The advantage of such a system is that it is easier for the statement's accredited environmental verifiers to prove that the information is correct and therefore reduces the time required on site and also the cost. It was also obvious, at an early stage, that everybody should receive basic environmental awareness training.

ENVIRONMENTAL POLICY

Since the inital writing of the environmental policy it has had several minor changes, but essentially the concepts of continual improvement and best practice have remained the same.

It is essential, when developing the policy, that all the site's senior managers "buy in" to it and this was achieved by them all jointly developing and agreeing the wording together. Although it is essential to BS 7750—and ISO 14001—that everyone understands the policy, it is clearly the senior managers who can carry it out effectively. The policy is:

> Akzo Nobel Chemicals Ltd, Site Gillingham has established and will maintain an environmental management system to fulfil the requirements of BS7750, 1994, as a minimum standard. This system, covering all the activities on site, will ensure compliance with current UK environmental legislation, best available practices and achieve a balance between economic, social and environmental responsibilities. We are committed to avoiding damage to the environment by any of our actions or operations.
>
> Site Gillingham is dedicated to continual improvement of environmental performance and efficient use of resources, which will be achieved by setting and ensuring successful implementation of environmental objectives. This policy will be made publicly available through the site annual environmental report and will be understood, implemented and maintained by all levels in the organization.

The environmental policy allows for the management to manage the business, especially to achieve a balance between economic, social and environmental responsibilities. It is clear that management plans to improve environmental performance as required by BS 7750 have different time scales and need to be reviewed regularly to ensure continual improvement.

TRAINING

After publishing the policy, all employees on the site received environmental awareness training. This half-day training session included general environmental awareness and explanation of BS 7750 and a group exercise to see how people could personally contribute to improving the site's environmental performance. The environmental awareness training took place in about 10 sessions on site. Feedback from these sessions was brought together into projects which were incorporated into the site's environmental targets and objectives. It is important at this stage that everybody on site is aware of the environment and feels part of managing the site's environmental performance.

Carrying out environmental awareness training at an early stage is essential. To ensure that any environmental management system works, everybody on site must be made aware of environmental issues, as a paper system will not work without the commitment of people. To reinforce the site's commitment to the environment, a site environmental magazine is published every six months to communicate performance and the progress of the site projects. It was interesting to note that the enthusiasm for environmental issues was much greater than with the original total quality training. This is further demonstrated with the setting up of an office recycling project. This project involves office personnel and has links with the local council and recyclers passing the message into the community, with some of the savings being given to charity.

MANAGEMENT RESPONSIBILITY

As with most major chemical companies, people are already responsible for environmental issues and responsibility is regarded as having equal importance to health and safety on the site. There are, however, prime environmental responsibilities where the health, safety and environmental manager is responsible, including legal waste disposal procedures and the maintenance of a register of current legislation. The operations manager is responsible for implementing environmental plant operations and the technical manager is responsible for monitoring environmental performance. These are clearly defined responsibilities. All senior managers have clearly defined environmental responsibilities and through the appraisal system all have at least one environmental goal.

LEGISLATIVE REGISTER

The site managers would not consider themselves to be experts in the area of legal regulations regarding its operations and therefore, to maintain an up-to-date system, they use the Barbour Index. This is a professionally prepared system which gives a quarterly review of updates or changes in legislation. The health, safety and environmental manager is responsible for ensuring that any changes to current legislation affecting the environment are communicated to the responsible managers on the site.

REGISTER OF SIGNIFICANT EFFECTS

The register of effects is the heart of any environmental management system. If this step in the implementation process is done well then the rest of the system becomes relatively easy. The procedure for the generation of the register of effects is, in principle, very simple. The first step is to generate an exhaustive list of all effects, then to assess their significance and finally to produce a register of significant effects.

When generating the exhaustive list of effects, consideration should be given to all raw materials, final products, processes, plants and buildings for their effect on air, water and land resource usage and nuisance, taking into consideration both normal and abnormal conditions. To produce this list could be quite a mammoth task; however, this information was already available through Control of Substances Hazardous to Health (COSHH) assessments. The COSHH assessment information was modified and put into a form which could be used to assess what the effects of the processes actually are, but not how significant they are. The site was again split into 25 areas and competent people were assigned to fill in the forms. This list was then screened against certain criteria which can define a significant effect. Deciding what is "significant" is not easy and cannot be based on one person's judgement. To make the system easy the following four criteria for significance were devised:

1. Where the magnitude of the effect, or the probability of its occurrence, is very small.
2. Where under normal or abnormal operating conditions there is a small probability of a minor environmental effect.
3. Where under normal or abnormal operating conditions there is a significant probability of a quantifiable environmental effect.
4. Where under normal operating conditions, the operation has the potential to cause a major effect on the environment or change the status of the site under environmental legislation.

Four different criteria for significance were chosen so that they ranged from an insignificant effect such as spilled water to a significant effect when under normal operations the plant or process is operating illegally. The exhaustive list of effects can then be given an environmental significance.

Each item in the effects register was assessed against these criteria and assigned a significance on the scale of 1 to 4. This was carried out by a team consisting of the health, safety and environmental manager, a representative of the production department and the process and quality control manager. It was decided that an effect of three or greater should be classed as a significant environmental effect and the register of significant effects could then be compiled.

Once this list of significant effects has been prepared, then, based on the environmental policy, targets and objectives for environmental improvement can be set. The list of significant effects produced a wide variety of objectives, but none of them a surprise. The cost implications were also varied. The projects ranged from an expensive £0.3 million waste water treatment plant to low-cost solutions such as rationalization of storage facilities and improvement of the site's emergency plan and training.

MANAGEMENT PROGRAMME

Having identified the significant effects, the site used a project management system to ensure that these environmental targets and objectives were met. Significant effects can be combined together to form individual projects which can be assigned different targets, objectives and time scales for implementation. The projects are reviewed on a quarterly basis by the site management team to ensure that target and objectives are being met and that the appropriate priority is assigned correctly. The site's environmental objectives can be summarized from the list of significant effects and from the original site environmental awareness training feedback.

The register of effects gives visibility to the problem areas and therefore allows them to be managed effectively.

SYSTEM INTEGRATION AND AUDITING

The site's quality management system and environmental management system structures are very similar and the requirements of BS 7750—and now ISO 14001—have been integrated into the site management system. This has many advantages: first, that internal audits are not carried out differently for ISO 9000 and BS7750. The site is also not auditing twice for the different standards. Internal auditor training is seen as an important part of maintaining the system, however the internal audits are basically ISO 9000 system audits. A management system internal auditor requires both ISO 9000 auditor training and practical environmental auditor training sessions for specific site needs. Although system auditing is all that is required by BS 7750, the next logical step is to carry out performance auditing, which will simplify the preparation of the EMAS statement.

Internal performance auditing and verification of the EMAS statement are closely linked. A verifer will spend less time verifying an EMAS statement if the internal performance audit system is adequate to meet requirements. Less time should mean less cost, so all can benefit from internal performance auditing.

EMAS REGISTRATION

Since the launch of the standard in early 1992, the site has implemented the environmental management system step by step and finally received accredited certification to BS 7750 in March 1994. The site is part of the polymer chemicals business unit, which is an international business, and therefore the next logical step was to go for EMAS registration and the final step was a verified environmental statement.

THE ENVIRONMENTAL STATEMENT

In 1994, the first site environmental report was produced based on the CEFIC guidelines. This quantifies the site's performance and makes the site policy and objectives publicly available. In June 1995 the second site environmental report was prepared, however, this was based on the EMAS guidelines and differed from the first report in 1994.

The first major difference was that the EMAS statement had to be site specific. Much of the corporate information included in the original report was not relevant to the environment. Second, it had to be written so that it was easily understandable by the public. This in itself is difficult, as many technical and legislative abbreviations are used and explanations had to be given. To obtain some unbiased comments, the statement was given to different non-technical people on the site, all of whom some very useful comments on the simplification and understanding of the environmental statement.

The advantage of providing a report in 1994 was that progress and achievements during the past year could be reported and also the future programme for 1995 and 1996 outlined.

The environmental performance indicators were related to the site's significant effects. However, guidelines on what actually need to be reported are, as yet, unclear. The report did include energy usage, environmental costs, waste disposal and fairly detailed information on waste water discharges. A statement regarding the use of raw materials was also required by the regulation and this caused the most problems. A measure of key raw material conversion efficiency was eventually produced but this is specific to the site and adopts a fairly technical approach.

THE BENEFITS OF EMAS REGISTRATION

Managing the environmental must be an integral part of any site's operation. As with health and safety, it does not guarantee that there will not be an incident but it does minimize the probability of one occurring. Managing the environment also makes very sound financial sense, in both the long and short term. An energy reduction project for the site showed a reduction of 18% from 1993 to 1994.

Fines for environmental pollution incidents are becoming larger and, along with the adverse publicity which a company receives, can significantly affect the future profitability and investment of a site.

It has taken three personyears to implement the environmental management system on the Gillingham site, without the use of consultants, and it has certainly been time well spent.

NOTE

It is acknowledged that the basis of this chapter has already been used in various materials by Akzo Nobel Chemicals Ltd, and in particular, as part of its EMS programmes.

Index

abatement technologies 222, 230
abnormal plant operation 162, 175–6, 321 344
accounting practices 82, 147
accreditation 23, 120, 133, 173, 184–5, 187
   certification 170, 211
   certification bodies 174
   criteria (AU/23) 177, 180
   organizations 120
   procedures 43
accredited certifcation 184, 187, 208, 338, 344
accredited environmental verifiers 133, 187, 340
activated sludge treatment 159
add-on-technology 219
administrative instruments 80
Agenda 21 19, 50, 52
aid orgnisations 56
air pollution 264
alternative technology 52
American National Standards Institute (ANSI) 29
approval system 232
Atmospheric Pollution Act 1965 244
audit 51, 66, 72, 88, 109, 115, 119, 121, 129, 168, 210, 248, 260, 343
   methodology 217, 260, 302, 343
   procedures 21, 211, 217, 256, 302, 326
auditing team/s 40, 185, 326
auditors 40–1, 43–4, 176, 180, 184–7, 301, 326
Australian Chamber of Manufacturers 118
Australian Environmental Protection Agency (EPA) 117
awareness raising 50, 130, 180, 239, 243, 260, 262, 269, 308, 340–1, 343

backwater technologies 92
Baldridge Award 29
Baltic states 272
banks 16, 164, 172, 274, 320
Barbour Index 342
barriers to trade 24, 31, 39,
best avaialble techniques 160–161
best available technology (BAT) 72, 79, 87, 140, 226, 232
best available technology not entailing excessive cost (BATNEEC) 160–3, 245–7
best environmental option 180
best practical environmental option (BPEO) 160–3
best practice 206, 312, 340
biochemical oxidation demand (BOD) 98
biodiversity 52
Blue Angel Programme 21
Brazilian Association of Technical Norms (ABNT) 261
Brazilian Congress 261
Brazilian legislature 261
bridging document 169
British Chemical Industries Association (CIA) 158–159
British Standard (BS) 5750 165, 168
British Standard (BS) 7750 23, 29, 37, 84–5, 87, 109, 115, 157, 165, 168–80, 188, 193, 199, 208, 213, 216–18, 234, 260, 295, 300, 308, 316, 318–19, 329–30, 335–6, 338, 340–4
British Standards Institution (BSI) 28, 30, 165, 168–9, 330
build-operate-transfer (BOT) business 91, 100
bureaucracy 67, 85, 133, 302
business associations 45, 118, 129, 132

Business Council for Sustainable Development (BCSD)   19
business management systems   116
buyers   38, 42

Canadian Chemical Producers Association   105
capability study   331
capacity building   251
capital investment   61
Central and Eastern Europe (CEE) countries   268, 270, 279
Central Europe   294
central government   29
centralized planning   301
centrally planned economy   269
centre for the Study of Financial Innovation (CSFI)   214
certificate/s   41
certification   4, 23–4, 34, 38–42, 44–5, 85–6, 88, 108–9, 121, 157, 165, 168, 170, 172, 183, 210, 261, 284, 292, 319, 321, 329–30, 335
   bodies   44, 121, 184–6, 208
   infrastructure   42–3, 46,
   institutions   43
   procedures   43, 84
   process   186
   system   41, 107
certified products   120
CFCs   148, 218, 232
chain-of-custody stewardship   67
Charter for the Environment (ICC)   29
Chemical Industries Association (CIA)   158–9, 206, 211
chemical industry   206, 208, 210
chemical oxygen demand (COD)   221–5
chemical waste   103
chemical waste treatment centre (CWTC)   103
Chemical Industries Association (CIA)   158–9
"Chief Inspector's Guidance Notes" (CIGN)   161
Chinese National Environmental Protection Agency   92
civil liability   298
cleaner fuel   263–4
cleaner processes   92, 102–3, 105, 160, 164, 260, 265

cleaner production (CP)   2, 4–9, 13, 18, 49–52, 54–6, 80, 86, 91, 93, 95, 102, 109–10, 118–20, 135, 144, 146, 162–3, 191, 193, 219–20, 229, 235, 241, 260, 279–80, 283–303
   CP assessment   56, 119, 291
   CP centre   53, 56
   CP economy   56
   CP methodology   301
   policy   50, 53
   procedure   54, 114–15
Cleaner Production Programme   3, 13, 49, 52, 219, 224, 226, 293, 303
cleaner production projects   92, 115, 221–2, 225, 284–5, 292, 298
cleaner production training   288
cleaner/clean technology   2, 4, 41, 45, 49, 51, 53, 55, 91–2, 102, 141, 158–9, 220, 223, 225, 231, 235, 291, 297, 327
clean-up costs   215
climate change   52
Clinton Administration   71
closed-loop material paths prinicple   79
Cold War   143
code of conduct   55
code of practice   251, 253
combined cycle gas turbine technology   163
command-and-control regulation   24, 77, 130, 286
command-and-control strategies   77–8, 143, 319
command-and-control system   69–70, 150
commercial advantage   216
Commission for Sustainable Development   50
Commonwealth of Independent States (CIS)   275, 278
communication procedure   333
competent body   86
competitive advantage   150, 152, 232, 235, 273
competitiveness   13, 16, 38–9, 51, 116, 145, 152, 193, 232, 284, 288, 320, 327
competence   185
compliance-base approach   152
compliance-based companies   69
computer-aided design   147
computer-based tools   146
computer-integrated enterprise   323
computerised database   49

INDEX

Confederation of British Industry (CBI) Environmental Business Forum 338
Confederation of Indian Industry (CII) 41, 45
conformity assessment 22–3, 43, 46, 292
conformity assessment infrastructure 42
consensus-based document 30
consultant/s 25, 40, 42, 44, 105, 221, 234–5, 295–6, 312, 324, 328, 342
consultant fees 42, 83
consumers 25–6, 35, 38, 81, 88, 93, 132, 230, 268
consumer group/s 25
consumer safety 52
contaminated land 175, 215, 268, 332, 336
continual improvement 24–5
continuous environmental improvements 5–6
continuous environmental performance 137, 140–1, 163–4, 289
contractors 139–40
Control of industrial major accident hazards (CIMAH) 308, 332
Control of Substances Hazardous to Health (COSHH) 175, 319, 342
control system 203
corporate barriers 68
corporate culture 288
corporate image 288
corporate performance indicator 217
corporate responsibility 262
corrective action 34
cost-benefit analysis 165
cost-effective certification 210
cost-effective strategy 71, 333
cost-effective business strategy 69
cost-effective investment 300
cost-efficient technology 199
costs 79
cost savings 307–8, 315
Council of Ministers 294
cradle-to-grave 92, 103, 176, 200
curricula 55
curriculum 44
customer/s 26, 28, 67, 73, 86, 88, 140–1, 149, 164–5, 170, 187, 235, 297, 320, 334
customer-driven production 149
customer loyalty 150
customer relations 144
customer satisfaction 27

customer requirements 107, 195, 208, 321, 322
Czech industry 240, 284
Czech Cleaner Production Centre (CPC) 284

Danish Environmental Protection Agency (EPA) 230
data management systems 69
decision-support tools 4, 196
decommissioning 245–6
Deming's plan-do-act cycle 208
demographic trends 70–1
demonstration projects 4–7, 50–54, 56, 92, 97, 100, 110, 115, 119, 219–20, 298
Deutsches Institut fur Normung eV (DIN) 30
developed countries 2, 37–8, 56
developing countries 2, 12–13, 20, 25, 37–8, 41–2, 44–6, 52, 54
document control 31, 114, 295
documentation 31, 114
draft international standard (DIS) 20–2, 30, 39, 120
dust prevention 244

Earth Day 143, 271
Earth Summit 19
Eastern Europe 267–8
eco-cycle principle 80
eco-design 117–18
eco-efficiency 149
eco-labelling 88, 98, 105, 132, 230
ecological planning 98
ecologically sustainable indistrial development (ESID) 92
ecological sustainability 229, 231
Eco-management and Audit Scheme (EMAS) 2, 7, 12, 24, 34, 37–8, 121, 129, 132, 137, 141, 157, 168, 170, 179, 187–8, 193, 199, 208, 216, 230, 261, 273, 279, 295, 307–8, 316, 318, 320, 324, 326–7, 329, 338, 340, 344
EMAS registration 133
economic advantage 220
economic analysis payback 295
economic benefits 285
economic development 52, 313

economic growth   52, 70, 101
economic incentives   54, 91, 95, 164
economic indicators   295
economic instrument/s   54–5, 79, 96
economic partners   184
economic reform   96
economic savings   289–300
economic tool   93, 95
economically viable application of best available technology (EVABAT)   140–1, 231
ecosystem/s   18, 72, 108, 175, 218
effects-based approach   108
emergencies   31, 114, 175–6, 210, 216–17, 322
"emergency situations rapid response" procedure   341
emission permits   81
emission standards   93, 234, 300
emission trading   73
employee involvement   4, 225, 233–4
EN 29 000 series   168
EN 45012   173
end-of-line requirements   230
end-of-pipe engineering   18
end-of-pipe facility   273
end-of-pipe investment   200, 204
end-of-pipe pollution   230
end-of-pipe solutions   1–2, 69, 200, 286, 300
end-of-pipe technology   4, 13, 102, 141, 193, 195, 202–3, 273, 285–7
end-of-pipe treatment   51, 41, 105, 284, 300
energy balance   295
energy consumption   206
energy crisis   268
energy efficiency   68, 70, 95
energy-intensive industry   99
energy resources   275
enforcement systems   91, 130, 143, 303
environment plan   56
environmental accounting   82, 147
  action programmes   230
  administration   272
  advisory panel   218, 334
  assessment procedure (self)   118
  assessment process   174
  advisers   86

agenda   144
audit   21, 29, 34, 82–4, 88, 109, 119 168, 170, 186, 193, 216, 260–2, 264, 295, 335
auditors   21, 23, 343
audit registration schemes   186
authorities   81–2, 195, 273, 279
awareness   78, 84, 113, 269, 273, 308, 340–1
benchmarking   226
benefit   2, 51, 115, 184, 288
capacity building   239, 243
challenges   153
charge system   279
Environmental Choice Programme   21
clean-up   146
code   78–9
communication   324
concerns   253, 272, 306
consent data   106
consultants   83
control   217, 261
competence   86, 185
costs   79–80, 131–2, 171–2, 344
criteria   39, 268
culture   327
data   340
damage   79
demands   225, 265
decision-making   73, 147
degradation   50, 79, 311
effects   108, 176, 180, 185, 225, 311, 314, 321, 323, 326, 334, 342–3
effects analysis   186
effects evaluation   176
effects register   175
engineers   146
externalities   69
fines   280, 345
"front runners"   272
fund   239, 246, 279
goals   4, 54,
groups   16, 24–5, 77
guide   149
hardware   70
hazards   217
hazard score   217
health   313
health officers   332
hotline   105

INDEX

impacts 2, 39, 56, 77–8, 87, 111, 114, 118, 139, 191, 195, 200, 204, 217, 250, 307, 312, 319, 321
impact assessment 82, 130, 195, 214
improvement 72, 121, 146–7, 149, 204, 225, 235, 287, 308, 347
incidents 318, 331, 345
indicators 73
information 148, 150, 152, 330, 332, 340
initiatives 81, 115, 165, 169, 330, 335
innovation 150–3, 208, 269, 273
insurance 215
issues 29, 45,137–9, 203, 218, 235, 260, 294, 308, 314, 317, 319, 322, 341
investigation 302
knowledge bases 147, 179, 330
labelling 19, 21, 29, 82, 88
labelling (EU) 130
law 65, 69, 71, 73, 78–9
law (New Zealand) 108
leadership 151
learning 151
legislation 1, 40–6, 130, 132, 137, 185, 296, 313, 324–5, 336, 338
liabilities 144, 172, 292
loads 224
management 18–20, 34, 44, 108–9, 144, 149, 218, 260, 264, 267, 270, 274–5, 278, 300, 305, 311, 316, 325, 334
management programme 31
management programme report (EMPR) 239, 244–8, 253
management standards 1–9, 21, 26, 85, 92, 135, 145, 169
environmental management system (EMS) 19, 23, 29–30, 34–5, 37, 39, 61–2, 71, 82, 86, 92, 109–10, 121, 129, 148, 158–9, 163, 165, 168–9, 180, 173, 199, 213, 218, 223, 225, 229, 235, 248, 271, 274, 278–9, 283–5, 287–289, 291, 293, 298, 306–7, 317–18, 320–5, 327–8, 343–5, 347
assessment 174
audit 34, 180, 187, 295
audit schedule 180
certification 23, 38, 109, 173, 185–7
certification bodies 120
guide 45
in the Promotion of Cleaner Processes and Products seminar 3–6
standards 12, 38, 45, 121, 184, 218, 293

tools 82
environmental managers 270, 274, 321, 323
mismanagement 269
objectives 39, 70–1, 111, 120, 345
objectives and targets 40, 12, 37, 40–1, 111, 113, 179, 343
opportunities 147
performance 18–20, 22, 25, 39, 66, 68–70, 73, 107, 118, 137, 139–40, 151, 159, 164–5, 170, 184, 186, 188, 193, 200, 204, 218, 220, 284, 291–3, 298, 305–6, 311–14, 317–18, 336, 342, 346
performance evaluation 21, 29
policy 31, 65, 67, 69, 71, 80–2, 111, 120, 137–8, 140, 143, 145–6, 158, 178, 248, 261, 279, 284, 321–2, 327, 342
politics 62
preservation 265
profiles 81
programme/s 42, 66, 73, 322, 325
Environmental Protection Act of 1990 (EPA 90) 158–9, 310
protection 2, 45, 80, 81–2, 119, 143, 158, 185, 235, 297
protection authorities 77
protection costs 86
quality standards 7, 41, 93, 133, 160, 162, 164, 318
questionnaire 179, 336
regeneration 314
regulations 40, 45, 61, 72–3, 93, 108, 119, 131, 151, 200, 225, 329, 331
requirements 39, 71, 115, 119, 210, 297
responsibility 39, 150, 261, 310, 342
results 71, 73
review 133, 314, 319, 321–3, 336
risk 172, 209, 214, 297
risk-management programme 213–14
risk-rating methodology 214
screening 214
services 235
significance 344
specialists 214
standards 17, 92, 165, 169, 173, 232
statement 323–4, 336, 346
status quo 152
strategies 82, 137, 225
support services 309, 311
targets 39–45, 345
taxes 80, 86

Environmental Technology Best Practice
    Programme   164
environmental values   73, 149
environmentalists   81
environmentally-friendly products   132, 263
environmentally-sound products   232
environmentally-sensitive operators   163
environmentally-sound technology   56, 70, 263
Euro-logo   34
European Accreditation of Certifciation
    (EAC)   188
European Commission   34, 135, 168, 188, 320
European Committee for Standardization
    (CEN)   22, 24, 28, 34, 135, 168–9
    mandate   135
European standards   34, 168
European Standards Organization
    (CEN)   135
European Union   12, 24, 34, 37–38, 160, 188, 199, 292–3, 317–18, 335
    eco-labelling scheme   88
    environmental policy   130–1, 201
    Fifth Action Programme on the
        Environment   129, 131–2
    legislation   178
    Phare programme   279
exporters   38
export earnings   38
export-led strategies   44
export-focused market   107
extended producer resonsibility   80
external assessors   180
external communication   113
external performance   183

facility audit   72
failure modes effects analysis (FMEA)   332
Fanie Botha Accord   244
Federal Register   66
Fifth Action Programme (EU)   129–32
fincancial aid   56
financial assistance   56, 119, 164
financial benefits   171
financial institutions   38, 81, 159, 164
financial services   101
financial support mechanism   132
first-party audit   23

first-party (internal) auditors   186
fiscal tools   73
fish-processing industry   219–20, 224
five-year plans   269
Ford Q1   328
foreign donors   279
foreign investers   297
Fourth Kondratiev cycle   268
full-cost accounting   67, 150
function-oriented legislation   294

General Agreement on Tariffs and Trade
    (GATT)   24, 292
general liability policies   215
generic model   165
Global Envrionemntal Management Initative
    (GEMI)   29
good practice   213
government agency   56, 71
government environmental programme/s   66
gradual pollution   215
green accounting   83, 230
green consumer   145
green goods   81
greenhouse gases   292
green management practices   141
green market   263
green planning   143
green protectionism   292
green taxes   230
green technology   74
gross domestic product (GNP)   101
ground contamination   294, 299
ground water contamination   115

hazard ranking   215
hazard score   217
hazardous waste   234
health and safety assessment   209
health and safety requirements   218, 233
health, safety and environmental (HS+E)
    disciplines   106
    management   205, 208
    performance   206, 211
    protection   292
    risks   210
    policy   105
Her Majesty's Inspectorate of Pollution
    (HMIP)   158–9, 161, 164
herring industry   219

horizontal standards 17, 20
horizontal supporting instruments 132
housekeeping (good) 223, 226, 235, 279, 288, 296
Hungarian Environmental Protection Act II (1976) 297
hydroesulphurization (MDT) units 263

incentive-based legislation 120
independent certification 183-4
independent third-party 173, 184
independent verification 210
index system 98
"Indicators of Performance'" (CIA) 206
industrial assoications 38, 50, 55, 118
  codes 206
  development 18
  ecology 143, 206
  economies 35
  efficiency 268
  engineering 69
  estates 102
  leadership 144
  nations 37-8, 140
  organizations 81, 118
  pollution 49
  processes 50, 158
  production 78
  responsibility 81
  sectors 50
  waste 102
industrialised countries 40-1, 44, 52, 54
industry-government research 147
information technology 198, 206
informative projects 219
innovation-friendly policy environment 152
innovative organizational structures 147
in-process technologyies 51
input/output approach 216
inspector 162
insurance companies 16, 38, 81, 84, 172, 215, 320
insurance premiums 86
integrated fuel gasification 163
integrated management systems 195-6, 199-200, 203, 210-11, 292, 315, 328
Integrated Pollution Prevention and Control (IPPC) 79, 226, 232
integrated pollution control (IPC) 158-60, 162, 294, 336
integrated quality and environmental management system (QEMS) 298-301, 328-9, 334
intellectual property rights 45
interactive workshops 250
interest rates 274
interested partes 113, 248, 253
internal audits 170, 174, 180, 186, 214, 320, 345
internal control (IC) regime 295
internal efficiency 183, 184
internal communication 113
internal risk control 215
International Accreditation Forum (IAF) 188
international agreements 79
International Chamber of Commerce (ICC) 29, 168
international community 73
International Electrotechnical Commission (IEC) 19-20
  International guidelines 43
  International legislation 214
  loan 56
  markets 52, 93
  obligations 160, 162, 164
  organizations 118
  regulations 51
  standards 12, 20, 25, 30, 39, 44, 108, 120, 133, 135, 165, 173, 184, 187, 239
  standards setting 40, 120
  trade 43
International Standardization Organization (ISO) 1, 6, 12 17, 19-20, 24, 28, 30, 35, 37-9, 119, 292
internet 119
inventory-type approach 216
investment decisions 51
  internal audits 23, 174, 180, 186
investment decisions 51
importers 41, 50, 55
Irish Standard IS 310 132
ISO member body 39
ISO draft international standard (DIS)
  14001 20-21, 39, 109, 120, 148, 336
  14004 20, 21, 120
  14010 21
  14011 121, 120
  14012 21, 120
  14060 21

ISO international standard
  ISO 14000 series   21–3, 25–6, 37, 84–5, 87, 186, 199, 260, 262, 273, 279, 283
  ISO 14001   2, 11–13, 23–4, 26, 30–1, 34, 37–9, 85, 109–11, 113–16, 120, 168, 170, 172, 174, 178, 180, 188, 193, 208, 213, 216, 231, 248, 283–4, 306, 314
  ISO 14004   23, 34
  ISO 9000 series   23, 25–6, 28, 38–9, 85, 87, 107, 109, 115, 133, 165, 168–70, 173, 180, 187, 193, 199, 208, 218, 283–4, 293–5, 298, 345
  ISO 9001   31, 208, 234, 319
  ISO 9002   306, 335
  ISO Technical Committee TC 176   17, 22, 28
  ISO Technical Committee TC 207   12, 17, 19–22, 24–5, 28, 39, 135
  TC 207 sub committee 1 SC1   29, 30, 31, 34, 135
  TC 207 sub committee 2 SC1   29
  TC 207 sub committee 3 SC1   29
  TC 207 sub committee 4 SC1   29
  TC 207 sub committee 5 SC1   29
  TC 207 sub committee 6 SC1   30
irrational prices   93

Joint Accreditation Scheme-Australia New Zealand (JASANZ)   120
just-in-time (JIT)   66, 180

Kaizen quality principles   234
Keidandran Principles   29

laboratory equipment   40
landfill site   329–30, 332, 334
Latin American Integration Association (ALADI)   43
lead compounds   263
lean and clean management   65–7, 70, 74
lean and clean production   67
lean production   66–7, 149
learning cell   147
legal compliance   213, 216
legal register   256
legal requirement   216, 248

legislative compliance   178, 305, 313, 316
legislative control   103
legislative requirements   102, 121, 291, 293, 299
lesser developed countries (LDC)   22, 292
liabilty   215
liason organization   20
liberal economy   232
liberalization   44
life cycle   78
  analysis   29
  assessement (LCA)   21–2, 67, 82–4, 105, 118, 150, 191, 193, 204, 292, 293
Lithuanian industry   240
living standards   231
local authorities   172
local communities   172, 310, 311, 320, 322, 325
local government   29, 99, 118, 311, 316
low-and-nonwaste technology   268
low-cost solutions   345

Maastricht treaty   292
management codes   253
  commitment   12, 41, 111, 113, 184, 284, 296, 301, 305
  control systems   206
  culture   172, 241
  plans   253
  practices   291
  problems   114
  review   34, 115–16
  systems   172, 206
  systems standards   17, 206, 210
  training   165
  transformation   306
manpower resources   169
marketable waste   220
market access   38
market-based economy   297, 299
market-based instruments   80, 129, 132
market-based solutions   80
market-based tools   132
market incentives   73
market mechansims   91, 96, 100
market profile   306
marketing opportunities   51, 293
material efficiency   149
matrix-based design tool   216
medium-sized companies   296, 327

INDEX

medium-sized manufacturing 235, 305
meterology equipement 40
mine closure 244–6
Minerals Act 1991 (South Africa) 244
Mines and Work Act (1956) 244
model 28, 109–10, 131, 165, 179
monitoring 114, 248
monopoly 259
monopoly supplier 161
multimedia solutions 72
multinational corporationss 178, 206
multistressor solutions 72

national accrediation bodies 43, 183, 187
National Environment Industries Database (NEID) 119
National Pollution Control Authority (NPCA) 295–6
national standards 39, 43, 133
national regulations 51
National Quality Award 260
natural gas 264
nature conservation 269
neoclassical economics 69–70
Netherlands Normalisatie Instituut (NNI) 29
New Zealand Chemical Industry Council (NZCIC) 109
New Zealand Resource Management Act 108
newly industrialized countries (NIC) 22, 25
new technology 41–2, 51–2, 232, 274, 296, 310
non-compliance 131, 158, 162, 176, 180, 186, 298, 316
non-conformance 34, 131
non-governmental organisation (NGO) 50, 164, 262, 320
non-tariff barriers 24
non-tariff trade barriers 37–8
normal operations 175
normative legislation 130–2
normative references 31
normative principles 81
North American Freee Trade Area (NAFTA) 292
North Sea Conference 164
Norwegian Cleaner Production Programme 295

Norwegian industry 294
Norwegian Technical Standards Institute (NTS) 30

occupational health and safety 16–17, 178, 210, 234
oil
  exploration 259
  industry 259
  products 259
  pollution 262–4
operational control 31, 177, 235
operational management 199
operational processes 114
opinion formers 206
optical industry 335
Organisation for Economic Co-operation and Development (OECD) 12, 275, 280
organizational change 151
organizational innovation 306, 319, 325
organizational performance 16
organizational procedures 319
OSPAR 164
output-focused system 31, 44, 50
overproduction 66
ozone 52

P-members (ISO) 39
payback period 97, 103, 110, 119, 232, 288, 332–3
perestroika 240, 272
performance-based system 31
permitting procedure 78
permit system 92
perscribed processes 158, 161
phosphahate-free detergents 70
physical prototypes 146
planned economy 273
plastics and rubber industry 335
policy fragmentation 151
policy implementation bodies 151
policy instuments 57
policy makers 144, 146
political instruments 195
polluters 93
polluter pays principle 79–80, 103
pollution abatement plant 160
pollution abatement technologies 158, 163
pollution control 50–1, 55, 92, 158, 296

pollution control technology   51–2, 55
pollution emmissions   78, 263
pollution fines   298
Pollution Prevention Pays programme   144
pollution prevention   12, 18, 66–9, 71, 111, 116, 217, 235, 280
pollution prevention strategies   62
pollution reduction   270, 272, 275, 278–80
population growth   231, 322
post-command-and-control paradigm   144, 152–3
poulation growth   70
power generation   72
precautionary principle   79–80
predictive modelling   146
preparatory environmental review (PER)   121
pressure groups   29
precautionary action   132
preventative action   132, 330
private sector   49, 66, 72, 74, 103, 150, 297, 311
private sector innovation   152
privatization   275, 297–8
process diagram   175
process improvement   51, 163
process innovations   193, 233–5, 321
process life cycle   152
process quality   25
process technologies   51, 162
producers   103, 105, 132
producer responsibility   80
product design   51, 53, 195
product chain   201
   differentiation   144
   innovation   232, 235, 319, 321
   life cycle   152, 193, 200
   planning   139
   quality   25, 288
   standards   21, 30
   stewardship   206
production efficiency   267
production costs   51
production process optimization   279
production technology   146
profitability   82, 198, 204
psychological barriers   44
public authorities   221, 318, 320
Public Environmental Reporting Initiative (PERI)   158–9
public images   51

public opinion   144
public policy   18, 66, 69–71, 73–4
public-private partnerships   151
public registers   160, 162
public reporting   4, 88, 274
public sector   49, 72, 74, 103, 145, 150
public sector organizations   144
purchasers   45, 81

Q-Base2   108
QEMS   298–301, 334
quality
   assurance   17, 84
   auditing team   328
   circles   234
   management   17, 109, 199, 292, 320, 324
   management systems   107, 115, 165, 199, 233, 328, 345
   management system standards   168
   management training   165
   manual   233, 295
   procedure   233
   standards   25
   system   42, 108, 187, 233, 295, 301
Quality System Assessment Recognition (QSAR)   43,

raw materials   79–80, 92–3, 110, 141, 147, 149, 175–6, 195, 200, 221, 232, 284, 288, 333, 344, 346
receptors   217
receptor score   217
recommendations   5–6
record keeping   114
recycling   54, 67, 79, 93, 97, 105, 201, 220, 225, 232, 264, 296, 316, 343
regional standard setting   39, 232
register of legislation   178
registration   23
registration costs   42, 54
regulated enterprises   131
regulators   16, 38, 131, 164, 172, 320
regulatory agencies   81
regulatory authorities   81–2, 86–7, 158, 225, 325
reglatory compliance   174, 186
regulatory framework   77, 115
regulatory policy   73
regulatory pressure   278

INDEX 357

regulatory requirements 71, 297
regulatory stategy 71
regulatory system 38, 151, 158, 163
regulatory system (US) 163
rehabilitation plan 244
remediation standards 115
resource efficiency 70–1, 175, 292
Resource Management Act (NZ) 108
resource maximization 68
resource management 208
Responsible Care programme (CIA) 105, 109, 158, 205, 208, 210–11, 336
reporting requirements 55
Rio Declaration 20, 79
risk assessment 82, 191, 193, 215–18
risk avoidance 214
risk-based environmental management 143, 213, 217
risk management 4, 214–15
risk profile 217
risk transfer 214–15

salesperson 28
sanctions 162
second-party (consulting) auditors 186
self-assessment 118, 210
self-certification 40, 42, 44–5
self-declaration 184
self-regulation 7, 129, 132
self-regulatory tools 4
service industry 149
shareholders 172, 188
significance 176, 180, 186, 213, 344
significant environmental aspects 113–14
significant environmental effects 306
significant impact 253
site landscape improvements 315
small and medium-sized enterprises (SMEs) 4, 12, 22–3, 30, 86, 107, 119, 121, 135, 139, 159, 234–5, 294, 305–7, 309, 310, 312–15, 318, 324–5
SME sector 305, 311
small scale techniques 6, 53
social responsibility 16
South Australian Cleaner Industries Demonstration Scheme 119
Soviet technology 299
Spanish Standard UNE 77–801 133
Spanish Standard UNE 77–802 133
specification-based document 34

stakeholders 56, 84, 184, 193, 206, 269, 279–80, 289, 307, 318, 320, 325
stakeholder requirements 13, 208
standard setting 43
standard-setting bodies 119
Standards Australia (SA) 29
standards bodies 28
standard making 292
standardization 43, 45–6
standardization bodies 38
Standards Australia/Standards New Zealand Committee 120
state-owned company 299
statuatory objectives 108
Strategic Advisory Group on the Environment (SAGE) 19
sudden and accidental reinsurance 215
sunset list 82
supercomputers 146
supplier/s 28, 39, 73, 81, 139–140, 148–9, 176, 179, 202, 320, 325, 332
  chain 148
  environmental assessment questionnaire 332, 334
  monopoly 161
  quotas 39
support measure 54
sustainable development 8, 13, 19, 29, 62, 78, 108, 120, 132, 143, 262, 307
sustainable management 108–9
sustainability 235, 305
strategic management 196, 199
strategic programmes 198–9
strategic planning 199
surveillance visits 174
Swedish Environmental Code 62, 78–9, 83
synthetic environment 146

tax/es 80, 275, 310
taxbase 311
taxpayers 230
tax rebates 55
tax systems 91, 278
teaching material 44
technical applications 49
technical assistance 43–4, 46, 55–6
technical barriers 42, 43, 46
Technical Barriers on Trade Agreement (TBT) 12, 38
technical equipement 41, 79, 91

technical knowledge   185
technical standards   12, 17, 165
technocratic approach   273
technological change   56
technological controls   193
technological cooperation   56
technological fix   220
technological innovation   71, 144, 306
technological investment   193
technological optimization   220
technological controls   193
technological restructuring   268
technological solutions   141, 318
technology
  advanced   91
  new   119, 232, 296, 310
  outdated   91–2
  policy   74
  pollution abatement   158
  transfer   44
third-party assessor   179
third-party audit   23
third-party audit (EMS)   186
third-party auditor (independent)   186
third party certification   39, 42, 45–6, 184, 206
third-party registration   21
top management   26
total management systems   191, 199
total quality management systems   66, 115, 204, 315, 319
total quality management programme (TQM)   68, 260
total quality strategy   306, 320
total quality training   343
tourism   101
Toxins Release Inventory   73
trade associations   28, 132
trade barriers   7, 12, 23, 37, 38–9, 45, 135
trade relations   39
trade unions   118, 311
trading partners   45, 46
training   31, 113, 141, 161–2, 179–80, 261, 299, 301, 318, 323, 328, 343
  courses   41
  material   53
  program   53
transition economies   7, 40–2, 44–5, 50, 55, 194, 239
transnational organizations   193

underwriter   215
unemployment   311
United Kingdom Accreditation Service (UKAS)   173, 186–8
United Kingdom Department of the Environment   135, 164
United Kingdom Department of Trade and Industry (DTI)   164
United Kingdom Secretary for the Environment   159–60
United Kingdom regulatory system   158
United Nations   50
United Nations Conference on Sustainable Development (UNCED)   52
United Nations Development Programme (UNDP)   12, 37
United Nations Environment Programme (UNEP)   3, 6,12, 18, 49–50, 53, 57, 92, 284
United Nations Environmental Programme (UNEP) Advisory Committee on Banking and the Environment   164
United Nations Conference on Human Environment   18, 261, 271
United Nations Conference on Environment and Development (UNCED)   19
United Nations Industrial Organisation (UNIDO)   38, 40, 42, 50, 53, 92, 283
United States Aid for Developments (AID)   279
United States Department of Energy   68
United States Environmental Protection Agency (EPA)   53, 71, 151, 293, 295
United Soviet Socialist Republic (USSR)   268, 271, 272, 280

value-added process   294, 300
value chain   147–9
Vehicle-Emission Control Laboratory   263
verification   133, 168, 295
vertical standards   17
Vienna Agreement   34
virtual prototyping   146–7
volatile organic compounds (VOCs)   72
voluntary aggreement   55, 80
voluntary approach   317
voluntary compliance   143
voluntary initiatives   206
voluntary sector   311

voluntary standards   12, 23–24, 37
voluntary standard setting process   46
voluntary systems   1, 141, 163–4

waste
  disposal   144
  disposal costs   169
  generation   296
  management, radioactive   158
  minimization   51–2, 68, 82, 86, 103, 105, 267, 275, 278–80, 314
  treatment   51, 43, 46
  water   93
Water Act 1956 (South Africa)   244

water conservation   93
water consumption   220, 222–3, 225
water pollution   244
worker productivity   68
workforce   310
working environment   288, 331
working methods   161, 165, 225
workshop   253
World Bank   92
World Commission on Enviornment and Development (WCED)   19
World Industry Council for the Environment (WICE)   153, 158–9
World Trade Orgainsation (WTO)   12, 22, 24, 38